Annual Meeting of the Society for Machinery Failure Prevention Technology (MFPT 2023)

Niagara Falls, New York, USA
26 – 28 July 2023

ISBN: 979-8-3313-0404-1

Printed from e-media with permission by:

Curran Associates, Inc.
57 Morehouse Lane
Red Hook, NY 12571

Some format issues inherent in the e-media version may also appear in this print version.

Printed by Curran Associates, Inc. (2024)

For permission requests, please contact Society for Machinery Failure Prevention Technology at the address below.

MFPT Society
715 Twining Road
Suite 135
Dresher PA 19025

Phone: (215) 346-2870
Fax: (215) 352-6902

https://mfpt.org

Additional copies of this publication are available from:

Curran Associates, Inc.
57 Morehouse Lane
Red Hook, NY 12571 USA
Phone: 845-758-0400
Fax: 845-758-2633
Email: curran@proceedings.com
Web: www.proceedings.com

TABLE OF CONTENTS

Rolling Element Bearings Analysis and Strategy Development in an Enterprise Using On-line Monitoring

This paper describes the emergent need to efficiently process and accurately characterize outlier or elevated acceleration data in a relatively young on-line vibration monitoring program for a commercial nuclear power generating fleet. The effort to do so is described for the specific case of anti-friction bearings. Good analysts do this efficiently and can quickly offer the conclusion that a degraded (or noisy) bearing is present. Accurately estimating the impact on machine reliability is more of a challenge. The premise of this effort is that an ongoing reliability assessment improves with the understanding of what the bearing is doing and the issue progression from the time identified. The objective of developing the accompanying presentation is to engage in discussions of how this is done and how it can be done better, and better...

This paper will attempt to describe things in a vendor neutral manner, avoiding commercial system or product names.

To describe the on-line vibration monitoring added: Accelerometers and some proximity probes and tach probes are installed on many pumps, fans, and compressors and of course the motors and turbines driving them. This is predominantly a wired, line powered system on a wired network system that is physically isolated from controls or process, but accessible via the internal business layer. The edge acquisition devices collect synchronous data for machine groups and are always scanning such that data acquisition on alarm with pre-trigger is supported. Parameters from the signals are developed both on the edge devices and on the servers if post processing is needed. Parameters are streamed to the existing enterprise data archive.

The backbone of the existing monitoring architecture is the enterprise data archive used with an automatic pattern recognition system [APR]. The APR includes clustering analysis supporting an aggregate model error across model parameters. Experience has been that aggregate model error is a useful tool for providing early warning of things changing warranting further review.

An initial look at many machine's signals showed elevated levels of acceleration, but levels of velocity that were not objectionable. With hundreds of machines, thousands of channels, the apparent variability of acceleration trends resulted in frequent aggregate error alarms in the APR models, resulting in the consideration to depreciate the influence of acceleration parameters. Better: we needed to learn what the acceleration data was telling us.

Objectives of this effort:

- Be able to track an issue with an anti-friction bearing.
  - Is it staying about the same? Or is it getting worse. Or have the indications started to accelerate and site notification with action recommended.
  - Track the issue accurately without the overhead of having to dig into the issue each time and update plots and trends.
- Look at a situation, with an unspecified bearing, but be able to make logical progress, understanding what is in the data and what bearing characteristics are indicated.
- Understand more about what is going on.
- Have a methodology to be able to analyze after a situation has played out and develop the pattern of degradation that could be a basis for prediction – process for continuing improvement.

The data available for analysis is in data sets and trends. As mentioned, trend data is streamed to the enterprise data archive, but on the servers associated with the on-line system both data sets and trends are available. Data sets are stored less frequently than the trend data and the data sets can be exported for analysis outside of the browser-based user interface. The browser-based interface is capable and is actively maintained and improved by the supplier, but the ability to access the data for whatever purpose is valuable and certainly a factor for the effort described in this paper to have been possible.

Housekeeping:

1X	Shaft Turning Frequency
FTF	Fundamental Train Frequency, the frequency at which a single ball or roller will return to the same position (this will be between 0.3 and roughly 0.45 of the 1X frequency)
BPFO	Ball Pass Frequency Outer race, the frequency that a dent in the outer race will be impacted by rollers passing over it
BPFI	Ball Pass Frequency Inner race, same as BPFO, but the dent is on the inner race which spins with the shaft
BSF	Ball Spin Frequency, the frequency of the rotation of the rolling elements themselves
Nr	The number of rollers in a single plane. A simple ball bearing has Nr balls. A spherical roller bearing found in a pillow block design typically has 2*Nr rollers, but Nr rollers in each raceway.
d	Rolling element diameter
D	Pitch diameter (diameter of circle containing the centers of the rolling elements)
α	Contact angle (= 0 for purely radial load of simple bearing; in which case the working geometry of the bearing can all be in one plane perpendicular to the shaft) In a deep groove ball bearing if the load has a component in the direction axial to the shaft the contact angle will increase as the axial load increases and the track of contact moves up the side of the groove.
Ψ	Used to represent the reoccurring combination = d*COS(α) / D, which is the effective roller diameter ratio to the pitch diameter. Note the use of asterisk '*' for multiplication will often be used throughout instead of implying or using 'x'

----TABLE 1---

FTF	$= [1X] * (1 - d * COS(\alpha) / D) / 2$ $= [1X] * (1 - \Psi) / 2$
Ψ	$= (d/D) * COS(\alpha)$ $= 1 - 2 * ([FTF] / [1X])$, useful as it is expected you can find [1X] and if a bearing is affecting the machine, then [FTF] is most likely found in an envelope spectrum. A peak at [1X] – [FTF] is also frequently seen.
BPFO	$= Nr * [1X] * (1 - d*COS(\alpha) / D) / 2$ $= Nr * [1X] *(1 - \Psi) / 2$ $= Nr * [FTF]$
BPFI	$= Nr * [1X] * (1 + d*COS(\alpha) / D) / 2$ $= Nr * [1X] * (1 + \Psi) / 2$
BSF	$= ([1X] * D / (2 * d)) * (1 - (d * COS(\alpha) / D)^2)$ $= ([1X] * D / (2 * d)) * (1 - \Psi^2)$

---TABLE 2---

Remarks:

In this paper the shaft turns, and the inner race turns as if fixed to the shaft. The outer race does not turn. Outer rings do turn, probably stepwise and occasionally, depending on effective friction. The rollers have a no-slip relation to both inner and outer races. This is also not universally true. And the contact angle and roller dimension apply to all rollers equally. These stipulations are not usually listed, and this is not a paper of rigorous mathematics. But each of these stipulations can be violated and if so, affect the signal. Inspect enough cases with complicated looking signals and spectral representations you will think about these possibilities. Occam's razor, or K.I.S.S. will be more the approach, but real situations are much more complicated than any first level 'digital twin'. This paper is about developing a digital twin and using algorithms to handle housekeeping (peak finding and matching predictions to peaks found) to be able to handle the complications in stride and perform monitoring and trending efficiently on a reasonably large scale.

Process walk-through:

Bearing known

1) Are you sure what bearing was actually installed? (Even though that is the only bearing stocked in the warehouse for that application, how sure can you be that that is the bearing in there?). Proceed, but acknowledge the possibility of things not being what you expect.

2) From recent searches of available information, the FTF, BPFO, BPFI and BSF are widely available as three decimal place ratios to [1X].

3) Do you know the contact angle for which these bearing frequency factors are given? Without specific knowledge, a working assumption is that the contact angle, α is zero for simple geometries and for more complicated geometries the nominal contact angle is often provided.

4) If available, pitch diameter, D and roller diameter, d are valuable data. D can be estimated in simple cases (simple or deep groove radial ball bearing) as the average of the outer ring OD and inner ring ID or bore. You might be able to deduce the roller diameter, d from given published FTF, BPFO, BPFI, BSF factors. For estimated D, which value of d results in the best match to the factors published? Again, keep in mind that a simple bearing digital twin is being built as a tool to deal with often complicated data. And not knowing the actual contact angle, knowing d/D puts you in the best position.

Bearing not known

1) What do you know?

   a. Roughly what is the diameter of the shaft?

   b. What does the bearing look like (in general)? Are there photos available – even if they are of the machine when a different bearing was installed, the relative geometry is useful. Remember, the ratio d/D is a key factor.

   c. Do you have a sense of the loading possible? Specifically: weights of horizontal rotors (not necessarily numbers but ideas); are belts used (belts generated high radial loads); is there axial loading? Consider the potential for thermal expansion to make bearings fight each other if the sliding bearing doesn't slide as easily as it should.

2) Speed of shaft [1X] is determined from a spectrum with pretty good resolution. A peak locate function is used for the best estimate. Pay attention to clues that the shaft speed may not be constant throughout the data set. Any spreading of the 1X peak might be such a clue. A vane pass or gear mesh value with known vane count or tooth count are good ways to get accurate shaft speed also.

3) Enveloping methods are used. This paper details an application built that uses both an 'Improved Envelope Spectrum'[(a)] and the more typical Envelope Spectrum[(b)], [IES] and [ES] respectively.

   a. Fast Computation of the Spectral Correlation , by Jerome Antoni, Ge Xin, and Nacer Hamzaoui of the Laboratoire Vibrations Acoustique at Univ Lyon. The authors made an algorithm available in the public domain for MatLab (see page 261 – noted this paper is from the August 2017 journal of Mechanical Systems and Signal Processing and the paper begins on page 248). The theory comes from the field of cyclo-stationary signal processing, which is used in the communications field.

   b. MatLab, with the Signal Processing Toolbox. Used is the 'envspectrum' function. MatLab help is well written and uses an antifriction bearing as an example.

4) Look for candidate FTF peaks in the envelope (or direct) spectrum. With knowledge of 1X and FTF the spectrum is reviewed for candidates for the FTF. The effective roller to pitch diameter,

Ψ = 1 – 2*([FTF] / [1X]) is used to visualize candidates. A survey of FTF factors from published bearing factor databases suggests that the range as a ratio to 1X between 0.3 and 0.48 is possible. But in terms of Ψ visualizing the aspect ratio of the bearing is helpful

A relatively high value (Ψ ~ 0.4) results in a bearing that looks like

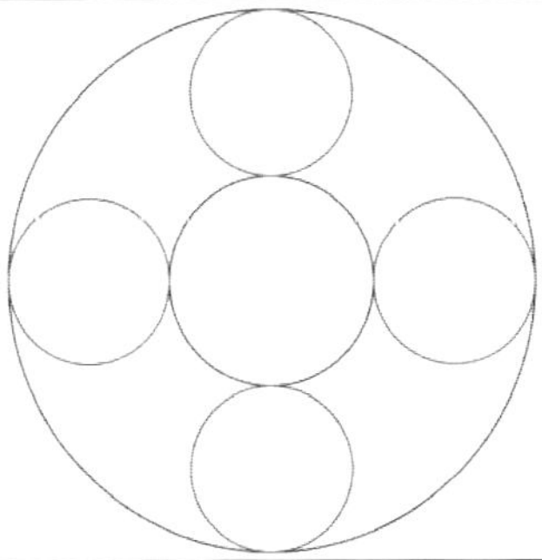

A lower value will look like

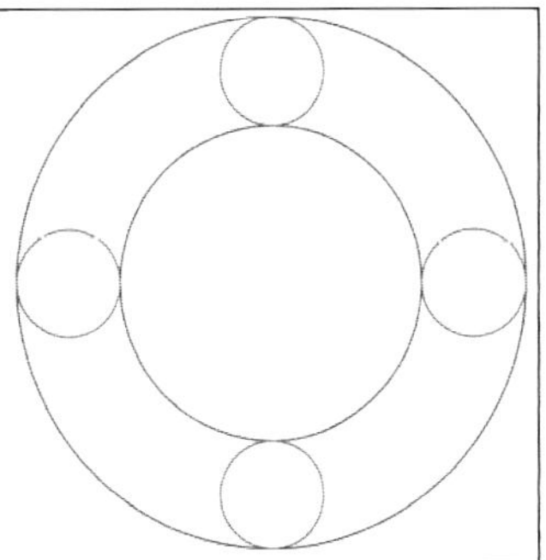

--- FIGURE 1---

This is where having a sense of the machine aspects helps.

a. Select the FTF from candidates (possibly work through the steps with more than one selected FTF). You need an FTF value to continue. (*The app used now allows a user provided value for FTF to be used so the analysis can proceed, not finding a candidate FTF in the data had occurred enough times to need this and similarly a user provided value of [1X] can be used).*

5) At this point either specify the number of rollers, Nr, or use a range that Nr will be within and you can look for BPFO and BPFI candidates. This can be done with a spreadsheet as BPFO is simply Nr*[FTF]. And using Ψ as noted above, BPFI is also calculated. A spreadsheet would calculate and display not only BPFO and BPFI for all Nr candidates (say in columns), but also multiples of each (in rows).

6) As with (5) above, but for BSF over potential contact angles, (α):

   a. For the candidate angle, α, and Ψ, determine d/D using the definition of Ψ.
   b. Then populate a matrix of possible BSF values for angles across columns and multiples of BSF in rows.

The math and use of a spreadsheet are not novel and not what this paper is about, but it serves to paint the picture of how straightforward the effort is conceptually. The application developed will:

- load the data set and identify a channel or channel pair (analytic signals are handled and provide good results when available – such as Horizontal and Vertical from the same bearing bracket).
- Help find 1X. Data sets from our system contain configuration information so a nominal value of 1X will exist (or take action to find the best estimate of 1X and enter manually). The spectrum near 1X is shown with 5 bin markers. Provided the center marker is on the highest amplitude bin value, the peak find works as intended and 1X is established.

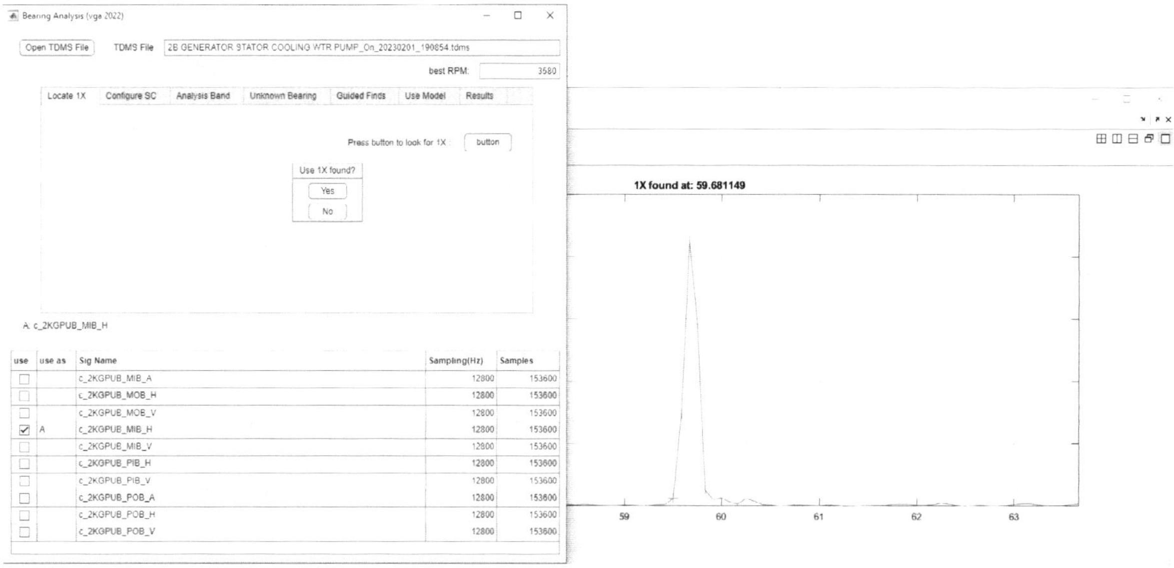

---FIGURE 2---

- Default settings for the [IES] developed from Spectral Coherence [SC] distribution are provided, but they are changeable. The [SC] and [IES] are calculated along with the traditional Envelope Spectrum, [ES] using the MatLab function with default settings. These are functional MatLab plots that allow you to zoom in and mark values. This functionality can be useful if for example you are wanting to understand what the spectrum looks like in the region where you should see FTF candidates. Envelope Spectra are often interesting.

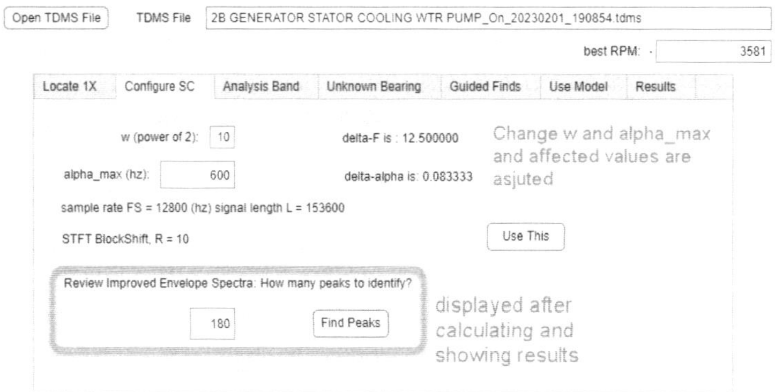

---FIGURE 3a---

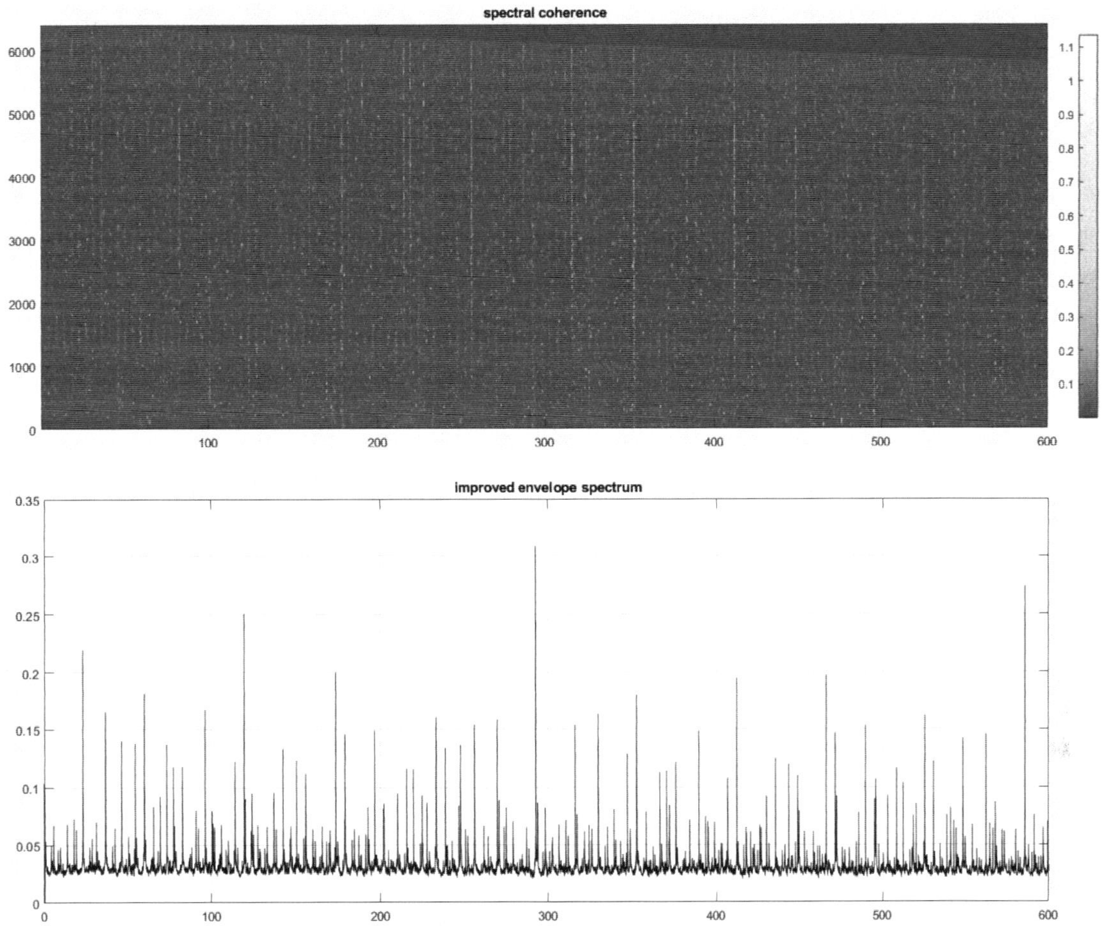

---FIGURE 3b---

The spectrum in Figure 3's lower plot is just an average value of the content of the distribution above, where the averaging is done vertically, so the frequency axis (X-axis) is maintained but the vertical axis content is averaged creating a single value for each X-axis value.

- The user is asked for the number of peaks to find in the Envelope Spectra. The default is a large number, discretion helps in some cases.
- In the next Tab the Analysis Band can be selected and the resulting IES and ES are plotted using the requested analysis band. And the user can choose the best band for continued analysis.

A note here about the Spectral Coherence distribution (the [SC]): think of this as starting with a spectrogram or time frequency distribution where the X-axis would be time and the Y-axis would be frequency content and using the color bar the amplitude at each point in time and for the frequencies from zero to Fmax are shown. Then conceptually take the spectrum of this spectrogram or time frequency distribution so that you have the Y-axis as before, but now the X-axis reflects the frequencies which are modulating various signal content components. Better yet read the paper cited associated with this work. Then, as different structural natural frequencies at the source or in the transmission path to the transducer will amplify different aspects of the modulation occurring, inspecting and using different Analysis Bands can take advantage of this.

Coherence has amplitude which represents a 'degree of coherence' obtained as fractional values from 0, for not coherent at all, to 1, for completely coherent. As a measure we would want something that tracks with the amplitude of the vibration associated with the issue to help with the judgement of severity. Therefore, the ES is calculated along with the IES. The IES offers clarity as to what is present, but the ES tracks with the issue in the units measured. The flow of the application at this point uses the IES to find things and create a good digital twin, and then the digital twin is used to identify the relevant content in the ES, or traditional envelope spectrum.

- Next the analysis begins:

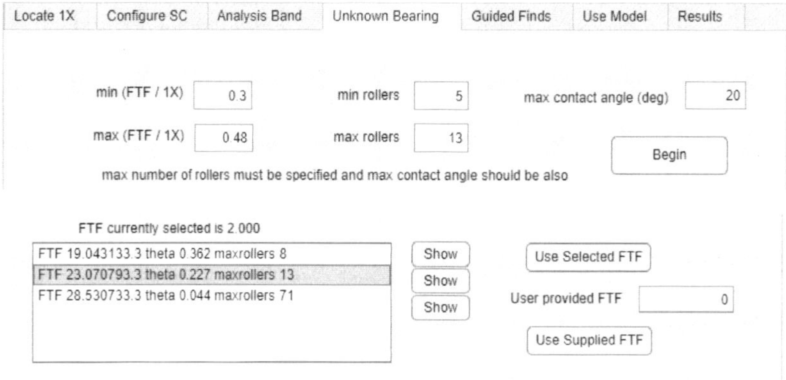

---FIGURE 4---

Three FTF candidates are found in the example followed and the Show buttons next to the candidates will show the graphics used above depicting the relative size of the rollers to the circles depicting the inner race and outer race to aid the selection.

With the selected FTF a pass is made comparing predicted FTF, BPFO, BPFI and BSF and their harmonics (along with 1X and n*1X) with the peak list generated from the IES. Results are presented in a table.

Initial results showing only direct hits and harmonics

Amp_order	Freq(hz)	Amplitude	ID_1	ERR_1	ID_2	ERR_2	ID_3	ERR_3
1	292.9455	0.4030	1xBPFIn8	0.0572	0	0	0	0
2	585.8903	0.3397	2xBPFIn8	0.1136	5xBSFa20	-0.2462	0	0
3	119.3755	0.2669	2x1X	0.0132	1xBSFa17	0.0758	0	0
5	23.0701	0.2629	1xFTF	0	0	0	0	0
8	329.5376	0.2402	1xBPFIn9	0.0383	0	0	0	0
11	489.5846	0.2244	4xBSFa11	-0.2501	0	0	0	0
15	256.3082	0.2211	1xBPFIn7	0.0309	0	0	0	0
21	471.9718	0.2040	4xBSFa19	0.1555	0	0	0	0
22	59.6755	0.2128	1x1X	-0.0056	0	0	0	0
24	549.2415	0.1931	3xBPFIn5	0.0759	0	0	0	0
25	238.7509	0.1851	4x1X	0.0263	2xBSFa17	0.1515	0	0
30	219.7104	0.1768	1xBPFIn6	0.0441	0	0	0	0
31	366.1795	0.1745	2xBPFIn5	0.0690	1xBPFIn10	0.0690	3xBSFa12	0.1057
36	179.0331	0.1685	3x1X	-0.0103	0	0	0	0
42	512.6548	0.1537	2xBPFIn7	0.1002	0	0	0	0
44	46.1407	0.1496	2xFTF	0.0005	0	0	0	0
50	495.0407	0.1317	4xBSFa7	-0.2425	0	0	0	0
60	246.8061	0.1192	2xBSFa8	-0.2672	0	0	0	0
67	358.1247	0.1126	6x1X	0.0379	3xBSFa17	0.2257	0	0
71	599.4093	0.1090	5xBSFa16	-0.1810	0	0	0	0
72	298.4042	0.1076	5x1X	-0.0015	0	0	0	0
104	247.8985	0.0869	2xBSFa6	-0.2361	2xBSFa7	0.2569	0	0
106	138.4168	0.0862	6xFTF	-0.0038	1xBPFOn6	-0.0038	0	0
107	69.2116	0.0860	3xFTF	0.0013	0	0	0	0
109	484.1093	0.0875	4xBSFa14	-0.0709	0	0	0	0
111	119.9663	0.0829	1xBSFa16	0.0482	0	0	0	0
112	123.4020	0.0826	1xBSFa8	-0.1346	1xBSFa9	0.1872	0	0
120	417.7756	0.0772	7x1X	0.0076	0	0	0	0

---FIGURE 5---

The peaks from the peak list are sorted by amplitude, largest to smallest and then in this order hits are displayed. The criterion for a hit is there is agreement within a definable error band. The default value for this error band is +/- 0.3 hz. This may seem generous, and it can be changed, and validation is possible later when the completed results are in hand and plots of the hits as cursors can be compared to spectra (IES and ES) directly (with the MatLab zoom capability). As the best angle is not yet identified a slightly generous error band is wise.

The sense of the error is (peak found – predicted) so a positive error such as the first value where the peak found at 292.9455 hz is 0.0572 hz higher than the prediction of 1xBPFO for a Nr = 8. Another error is associated with peak found at 247.8985 hz being 0.2361 hz less than the prediction of 2xBSF for an angle of 6 degrees (Amp_order value 104). Tracking this issue, we know that this is a relevant hit, but from the error values at 6 degrees (-0.2361) and at 7 degrees (0.2569) it is expected that the best angle will be something like 6.4 degrees. And based on this sensitivity of BSF to angle, the selection of an error band of +/- 0.3 hz seems a good one.

A note regarding BSF: 2xBSF and 4xBSF are given more credence than 1x or 3x if BSF is felt to be due to defects. This because a pit on a roller will likely contact both outer race and inner race in one rotation.

However, in this work (especially it seems with motors that have the same bearing at either end) noisy bearings have been identified where the only issue seems to be BSF with side band families, and the amplitude comes and goes as well with the contact angle changing. This is felt to be a result of temperature fluctuation and the ability of the accommodating bearing (the one intended to slide axially in the bracket fit) to actually slide. In these situations, the picture of the BSF and side band interactions with other things suggests that there are sweet spots where many things line up and the dynamic system simply has local maxima of noisiness at these sweet spots due to positive feedback.

- Continue analysis with application Tab: 'Guided Finds'

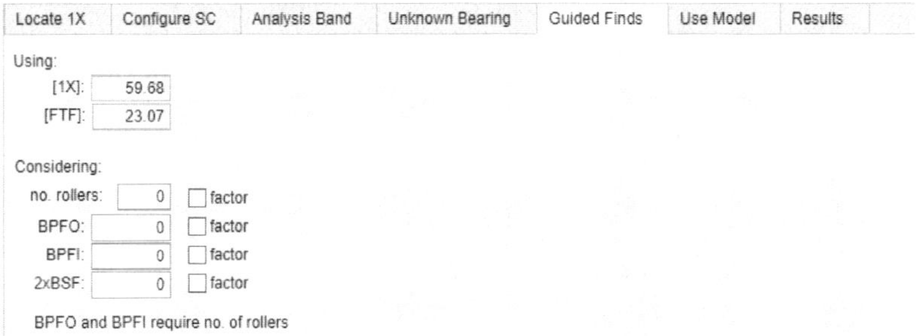

Use the Initial Results listing of peaks identified to fill in frequencies identified and ensure 'factor' check boxes are checked.

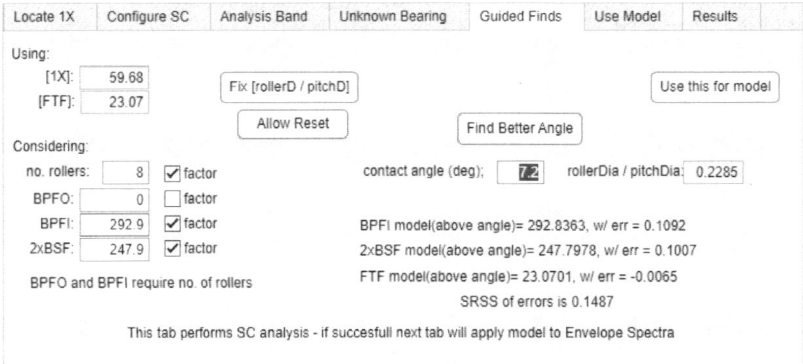

---FIGURE 6---

- Cut and paste frequencies from listing of peaks identified and fill in Nr, BPFI and 2xBSF
- Contact angle edit field comes up with 0, replace with 6.4 from above discussion (error values suggesting between 6 and 7). Noted is rollerDia/pitchDia edit field showed 0.2284 – previous work resulted in evidence that this should 0.2285 (known bearing vs unknown bearing is blurred hereon as the distinction is not really needed with the intent being to outline the process with glimpses of the detail on the way to automation efforts). Prior to pressing Fix[rollerD/pitchD] button, adjust contact angle upward until 0.2285 reached (6.75 works)
- Push button 'Fix[rollerD/pitchD]' and adjust contact angle to a value that seems to strike a good selection considering error associated with BPFI and 2xBSF. Noting that the error associated with FTF changes also.

Here it is that the digital twin is being exercised. Fixing d/D and then selecting α establishes the model and then the α parameter is used to fit the data.

- In this case the angle 7.2 seems a good choice. The 'Use for the Model button is pressed', and the listing below is shown:

Identified peaks from guided analysis with contact angle 7.20

Amp_order	Freq(hz)	Amplitude	ID_1	ERR_1	ID_2	ERR_2
1	292.9455	0.4030	1xBPFIn8	0.1092	0	0
2	585.8903	0.3397	2xBPFIn8	0.2176	0	0
3	119.3755	0.2669	2x1X	0.0132	0	0
4	412.3204	0.2654	1BPFI:1Xrsb_2	0.1218	0	0
5	23.0701	0.2629	1xFTF	-0.0065	0	0
6	173.5696	0.2465	1BPFI:1Xlsb_2	0.0956	0	0
7	466.5149	0.2408	2BPFI:1Xlsb_2	0.2046	0	0
9	96.3066	0.2297	2x1X:FTFlsb_1	0.0209	0	0
10	269.8755	0.2291	1BPFI:FTFlsb_1	0.1158	0	0
12	36.6148	0.2378	1x1X:FTFlsb_1	0.0103	0	0
13	352.6039	0.2219	1BPFI:1Xrsb_1	0.0864	0	0
14	316.0150	0.2151	1BPFI:FTFrsb_1	0.1021	0	0
16	526.1741	0.2156	2BPFI:1Xlsb_1	0.1826	0	0
19	562.8208	0.2113	2BPFI:FTFlsb_1	0.2248	0	0
20	233.2426	0.2136	1BPFI:1Xlsb_1	0.0874	0	0
21	471.9718	0.2040	1BPFI:1Xrsb_3	0.0920	0	0
22	59.6755	0.2128	1x1X	-0.0056	0	0
23	54.1927	0.1919	4x1X:FTFlsb_8	0.0810	1BPFI:1Xlsb_4	0.0810
25	238.7509	0.1851	4x1X	0.0263	0	0
26	73.2351	0.1850	2x1X:FTFlsb_2	0.0260	0	0
27	347.1377	0.1831	2BPFI:1Xlsb_4	0.1897	0	0
28	531.6939	0.1830	1BPFI:1Xrsb_4	0.1330	0	0
32	150.4996	0.1708	6x1X:FTFlsb_9	0.1021	0	0
36	179.0331	0.1685	3x1X	-0.0103	0	0
40	142.4450	0.1628	2x1X:FTFrsb_1	0.0061	0	0
44	46.1407	0.1496	2xFTF	-0.0125	0	0
45	82.7383	0.1531	1x1X:FTFrsb_1	-0.0195	0	0
46	113.9001	0.1481	5x1X:FTFlsb_8	0.1072	1BPFI:1Xlsb_3	0.1072

---FIGURE 7---

The highest 7 peaks are found, and note the scroll bar and size, suggesting the full list of peaks found is about twice what is shown in this window. This example data set is indeed a rich one that has been well studied.

At this point the digital twin with d/D = 0.2285 and α = 7.2 degrees is identified as a good fit to the SC, IES data peak set.

- Use identified configuration to address traditional Envelope Spectrum. Initial results for direct hits (1X and bearing frequencies with their harmonic families) found in Envelope Spectrum:

Envelope Spectrum results showing only direct hits and harmonics

Amp_order	Freq(hz)	Amplitude	ID_1	ERR_1
1	292.9172	0.1081	1xBPFIn8	0.0809
3	59.6663	0.0744	1x1X	-0.0149
9	119.3709	0.0508	2x1X	0.0086
18	238.7503	0.0346	4x1X	0.0257
31	247.9159	0.0258	2xBSFang*	0.1181
38	358.0814	0.0212	6x1X	-0.0055
39	123.9180	0.0211	1xBSFang*	0.0191
42	179.0772	0.0193	3x1X	0.0337
50	23.0632	0.0188	1xFTF	-0.0134
74	585.8724	0.0151	2xBPFIn8	0.1997

---FIGURE 8---

This is a smaller list but identified are 1xBPFIn8 (shorthand for BPFI for 8 rollers), 2xBSFang* (shorthand for 2 times BSF for the specified angle – in this case the 7.2 degrees) based on the configuration from the preceding SC, IES analysis.

Also, at this point, note that frequencies are not the same between direct spectrum, IES and traditional envelope spectrum. In this data set analysis for example:

1X from direct spectrum	59.681 hz
1X from IES (3000-5800 hz)	59.676
1X from envelope spectrum (3000-5800 hz)	59.666

This is typical, for 1X, and other frequencies, and for analysis performed on other data sets for this and other bearings.

- Complete the envelope spectra analysis by using found frequencies in edit fields and if desired tweaking the angle (d/D is now held fixed as the digital twin is established)

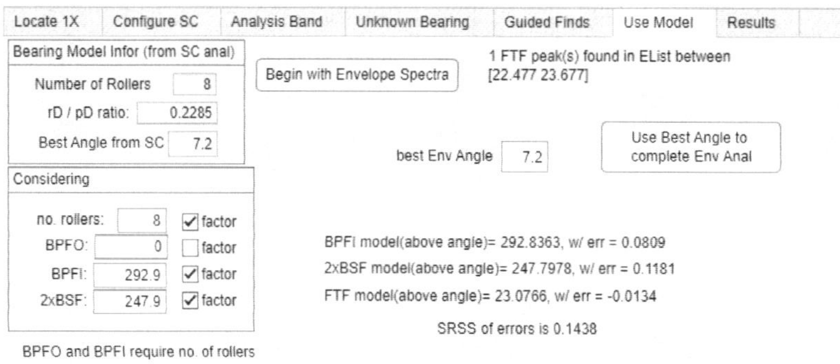

---FIGURE 9---

The error is reduced for BPFI by lowering α and the error for 2xBSF and FTF is reduced by raising α, which is understood by looking at the form of the for expressions for these frequencies (different signs on terms containing cos(α)). The following table is developed when 'Use Best Angle to complete Env Anal' button is pressed.

Envelope Spectrum results for contact angle 7.10

Amp_order	Freq(hz)	Amplitude	ID_1	ERR_1	ID_2	ERR_2	ID_3	ERR_3
1	292.9172	0.1081	1xBPFIn8	0.0809	0	0	0	0
2	36.5948	0.0763	1x1X:FTFlsb_1	-0.0231	0	0	0	0
3	59.6663	0.0744	1x1X	-0.0149	0	0	0	0
4	227.7516	0.0697	2BPFI:1Xlsb_6	0.1658	0	0	0	0
7	531.6684	0.0516	1BPFI:1Xrsb_4	0.1075	0	0	0	0
8	352.6085	0.0499	1BPFI:1Xrsb_1	0.0910	0	0	0	0
9	119.3709	0.0508	2x1X	0.0086	0	0	0	0
10	412.3342	0.0472	1BPFI:1Xrsb_2	0.1356	0	0	0	0
15	96.3109	0.0375	2x1X:FTFlsb_1	0.0118	0	0	0	0
16	12.3317	0.0358	6x1X:FTFlsb_15	0.1930	0	0	0	0
18	238.7503	0.0346	4x1X	0.0257	0	0	0	0
22	580.4116	0.0315	2x1X:FTFrsb_20	-0.2148	0	0	0	0
23	82.7444	0.0312	1x1X:FTFrsb_1	0.0000	0	0	0	0
24	233.2527	0.0308	1BPFI:1Xlsb_1	0.0975	0	0	0	0
26	173.5829	0.0304	1BPFI:1Xlsb_2	0.1088	0	0	0	0
27	471.9914	0.0302	1BPFI:1Xrsb_3	0.1117	0	0	0	0
30	73.2462	0.0264	2x1X:FTFlsb_2	0.0103	0	0	0	0
31	247.9159	0.0258	2xBSFang*	0.1181	0	0	0	0
32	89.4260	0.0246	10x1X:FTFlsb_...	0.0051	0	0	0	0
33	347.1646	0.0240	2BPFI:1Xlsb_4	0.2165	0	0	0	0
35	557.3340	0.0224	2x1X:FTFrsb_19	-0.2292	0	0	0	0
38	358.0814	0.0212	6x1X	-0.0055	0	0	0	0
39	123.9180	0.0211	1xBSFang*	0.0191	0	0	0	0
41	526.1907	0.0203	2BPFI:1Xlsb_1	0.1992	0	0	0	0
42	179.0772	0.0193	3x1X	0.0337	0	0	0	0
43	156.0026	0.0190	3x1X:FTFlsb_1	0.0224	0	0	0	0
46	48.9256	0.0188	7x1X:FTFlsb_16	0.1689	0	0	0	0
48	321.5056	0.0184	5x1X:FTFrsb_1	0.0366	0	0	0	0

---FIGURE 10---

Again, lots of frequencies found with most of the 10 highest peaks identified. To sum energies, peak values of finds are aggregated on a square root of the sum of square's basis. The shorthand example is 1X_FTF_sb means FTF side bands on either side of 1X and of harmonics of 1X.

Element	SC, IES result	Env spec result
FTF	0.0188	0.3261
1X	0.1007	0.4576
1X_FTF sb	0.1254	0.6862
BPFO	0	0
BPFO_FTF sb	0	0
BPFI	0.1091	0.5271
BPFI_1X sb	0.2157	0.9105
BPFI_FTF sb	0.0926	0.6116
BSF	0.0333	0.1218
BSF_1X sb	0	0
BSF_FTF sb	0.1518	0.5271

---TABLE 3---

In addition to the above features data set ID, channel(s) used, the analysis band used, 1X (from direct spectra), d/D, α (used for both IES and env spectra analysis), full peak lists with identified hits captured for both IES and env spectra analysis are saved for the analysis performed.

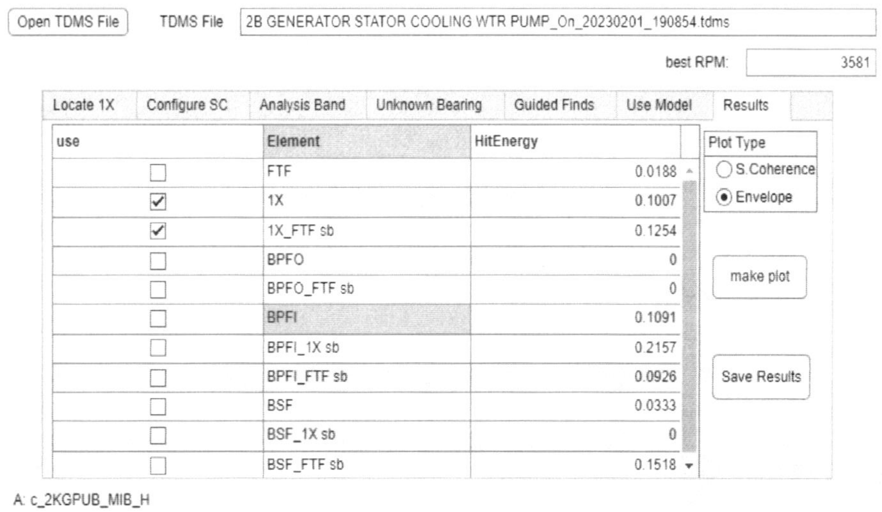

The results Tab shows elements from IES and ES analysis, allows plots to be made and supports saving the Results

The plot selected with these selections is of the traditional envelope spectrum, the ES, with 1X and harmonics marked in red and FTF side bands in blue.

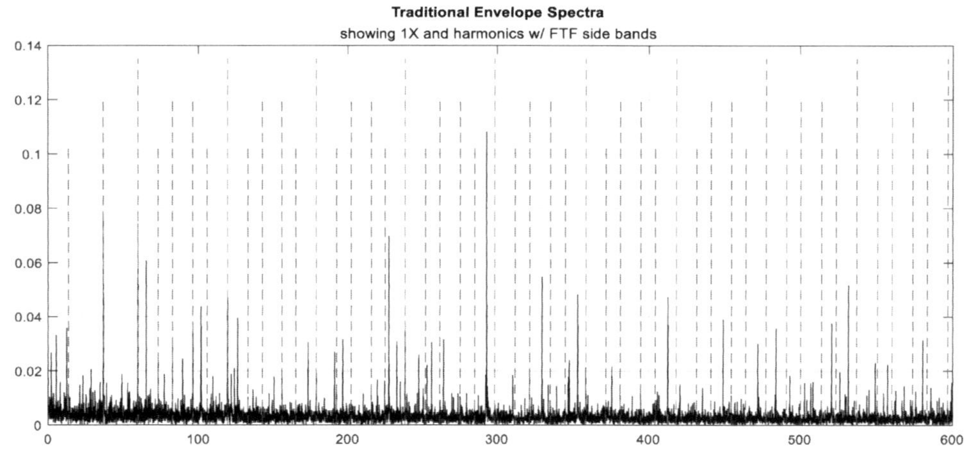

---FIGURE 11---

Side bands use a decreasing height shoulder to help make sense of busy plots. Noted is that seeing FTF reflected off 1X is a common finding ([1X] – [FTF]).

Below, BPFI w/ 1X and FTF side bands is shown.

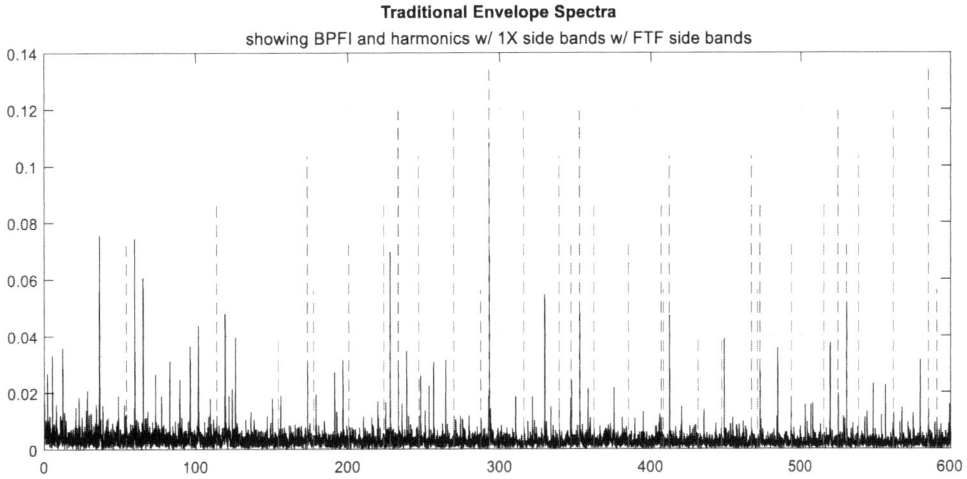

Same plot zooming in to right side of 1xBPFI

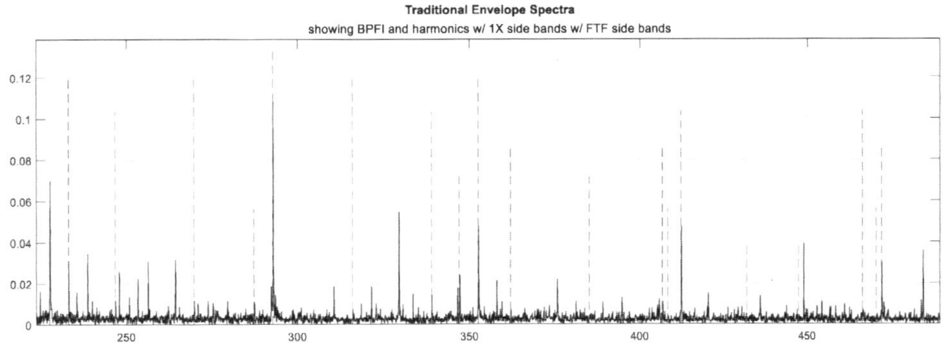

---FIGURE 12---

BPFI w/ side bands from SC, IES analysis shown:

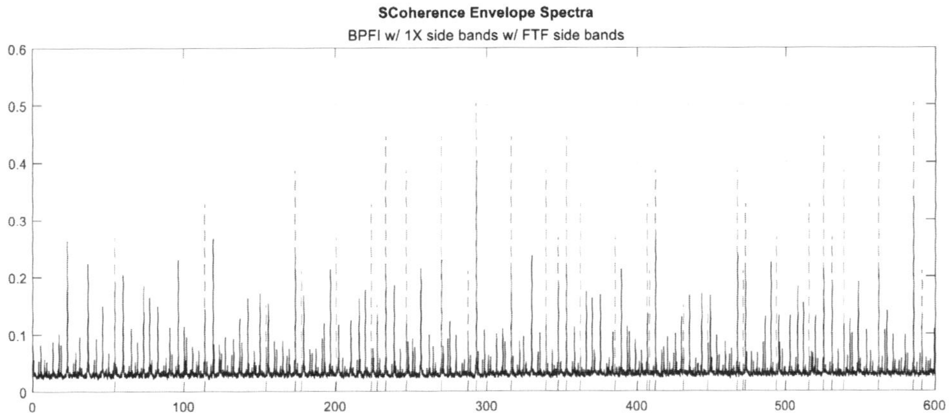

---FIGURE 13---

In fig. 13, note that the floor is not at zero amplitude. This reflects the unusual nature of this spectra, but the clarity of things found makes this a very useful analysis tool.

Below, zooming into to right side of BPFI as above:

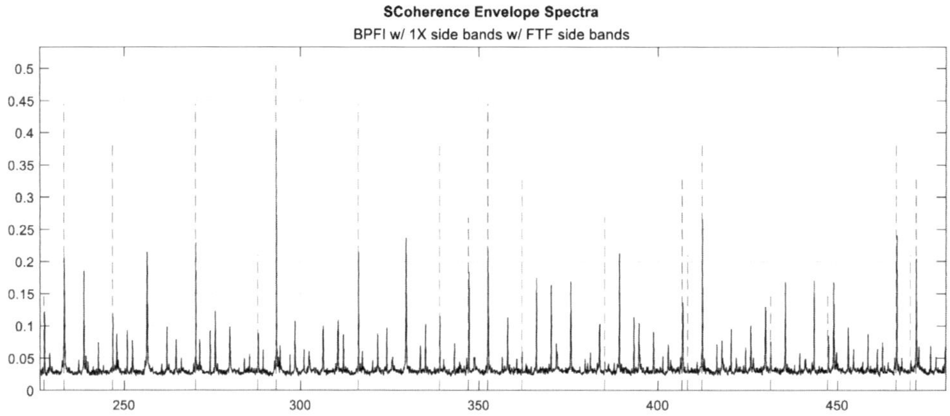

---FIGURE 14---

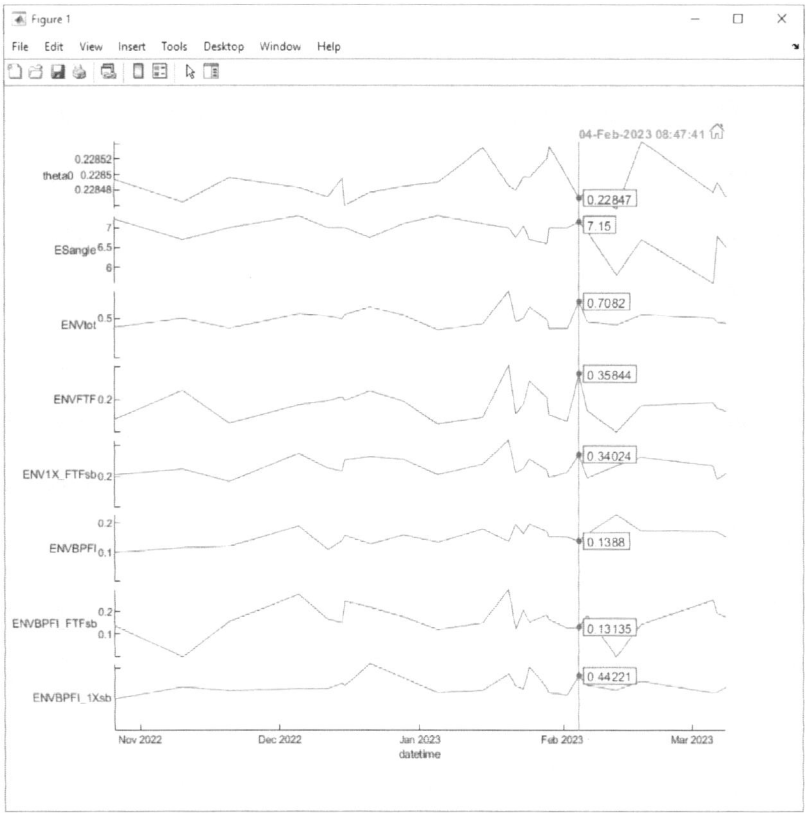

Hit energy parameters can be saved. This is a simple stack plot of some parameters saved from the manual analysis performed using the app.

---FIGURE 15---

The duration of the run through takes about 5 to 10 minutes (which includes a little time looking at plots). The FastSC calculation of Spectral Coherence depends on the size of time samples and value of 'w' which sets size of the window used in the short time spectra. A four second data set sampled at 12800 hz (our typical setting for points on antifriction bearings – noting the probe mounting method described limits the added value of sampling faster) and a 'w' value of 10 will take about 20 seconds. A 12 second data set with a 'w' value of 10 takes about a minute. These estimates based on experience with a fairly capable computer.

The analysis process aided by the app for at least one, but likely a few data sets for a situation, helps develop a sense of how that analysis progresses. The experience is that different issues suggest different analysis approaches. Should there be clarity as to a BPFI issue or a BPFO issue being dominant, then the clear approach is to have the main thrust to be a directed tracking of the specific target (BPFI or BPFO) first in the spectral coherence based IES as the data clarity is best and then with the digital twin to re-find that target in the normal envelope spectra, ES and obtain the tracking metrics.

Automating the scripts in some cases proved to be easy, but that was a trap. As an Engineer who doesn't always practice the best coding (code is documented as breadcrumbs are needed to pick up the effort after an interlude for example, but sometimes reckless assumptions are made that things will always follow possible branches in possibilities). It didn't take long to realize that addressing just about all possible branches of possibilities had to be covered for a script to be reliable.

The current state of the effort is that we intend to purchase a deployable engine for these scripts to run as services – likely acting when new unprocessed data sets are added to a queue or input folder to process the data, and add to a table of results that currently look like this abbreviated version:

datetime	Angle	ENVtot	ENVFTF	ENV1X	ENV1X_FTFsb	ENVBPFO	ENVBPFO_FTFsb	ENVBPFI	ENVBPFI_1Xsb	ENVBPFI_FTFsb	ENVBSF	ENVBSF_1Xsb	ENVBSF_FTFsb
4/11/2023 0:00	8.2117729	0.6181776	0.080020409	0.1750656	0.448010359	0.448010359	0.038680136	0.351909832	0.562195478	0.1961797	0.044615342	0.363477854	0.391769668
4/11/2023 2:00	11.494265	0.692546249	0.083614623	0.155360355	0.3600537	0.3600537	0.083561178	0.473766676	0.648006643	0.21539536	0	0	0
4/11/2023 4:00	12.992238	0.770574638	0.114333361	0.151005804	0.494004521	0.494004521	0.094602858	0.535997678	0.685147579	0.313823746	0.026036184	0.200014621	0.62271933
4/11/2023 6:00	8.1490755	0.798360621	0.103212372	0.167106953	0.476656381	0.476656381	0.064581142	0.608019026	0.711775649	0.359653796	0.027058277	0.174649309	0.703162266
4/11/2023 8:00	11.011619	0.773288368	0.047506767	0.169734086	0.408624022	0.408624022	0.052438015	0.580908391	0.709964256	0.240211117	0.062523887	0.593480434	0.351167374
4/11/2023 10:00	12.427206	0.802908997	0.054143684	0.181628282	0.698985096	0.698985096	0.040589078	0.63067099	0.779133003	0.204807833	0.016978588	0	0.717710076
4/11/2023 12:00	12.452539	0.965304905	0.08268877	0.201599887	0.380810232	0.380810232	0.076510305	0.793373838	0.935511417	0.236217369	0.035476061	0.529112406	0.737581963
4/11/2023 14:00	10.427185	1.074636956	0.08138382	0.218851535	0.293756649	0.293756649	0.093488281	0.883456062	1.021538797	0.321779096	0	0	0
4/11/2023 16:00	12.70547	0.986999918	0.094976814	0.240699889	0.478907944	0.478907944	0.054772509	0.776308171	0.924222638	0.330106188	0.043992971	0.366116417	0.845220148
4/11/2023 18:00	12.445043	0.914799082	0.103319668	0.238084589	0.715839811	0.715839811	0.072581284	0.705415918	0.846880727	0.296386322	0.031663977	0.298481256	0.685350647
4/11/2023 20:00	13.13575	1.056583052	0.232091831	0.277543652	0.47339607	0.47339607	0	0.867320869	0.976310832	0.399487899	0	0	0
4/11/2023 22:00	10.21946	0.957381374	0.087199188	0.273662177	0.339433703	0.339433703	0.095909113	0.75771577	0.846419339	0.420070035	0.025576534	0.310830072	0.773225773
4/12/2023 0:00	13.348962	0.940271277	0.256470902	0.199217024	0.412583846	0.412583846	0.084596523	0.753072586	0.893075648	0.249119392	0.028359785	0.173596653	0.882621158

---TABLE 4---

Processing a list of 100 data sets (roughly 4 MB each) and making this table takes about 20 minutes. Currently there are on the order of 10 actual 'interesting' bearings identified, so consider a planning value of 40 possible concurrent issues and using 30 seconds as the working value for a single data set to be processed, then a serial throughput of 40 within the frequency of data set creation is needed to keep up. Our standard data set capture frequency is 1 hour. Then using a basis of 2 data sets / minute, if data sets were being captured at 20-minute intervals, a serial estimate suggests this is the point of keeping up.

This is workable.

Case History:

A large vertical pump [design point: 5100 GPM / hot water specific gravity 0.873 / 875 ft. total head] of a vertical can pump design. Lower Seal housing to discharge head leak 'repaired' with a stop leak product injection. Seal starts leaking (steaming) in December. The motor IB Y channel acceleration (highlighted parameter) is spotted trending up before anything else:

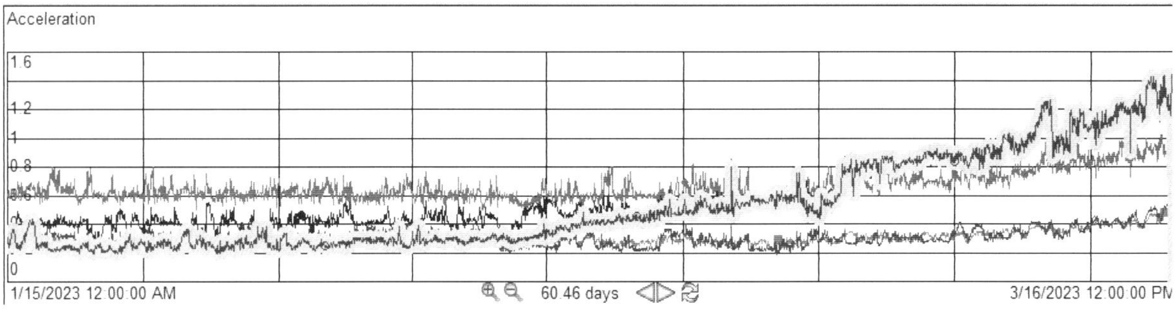

Other parameters:

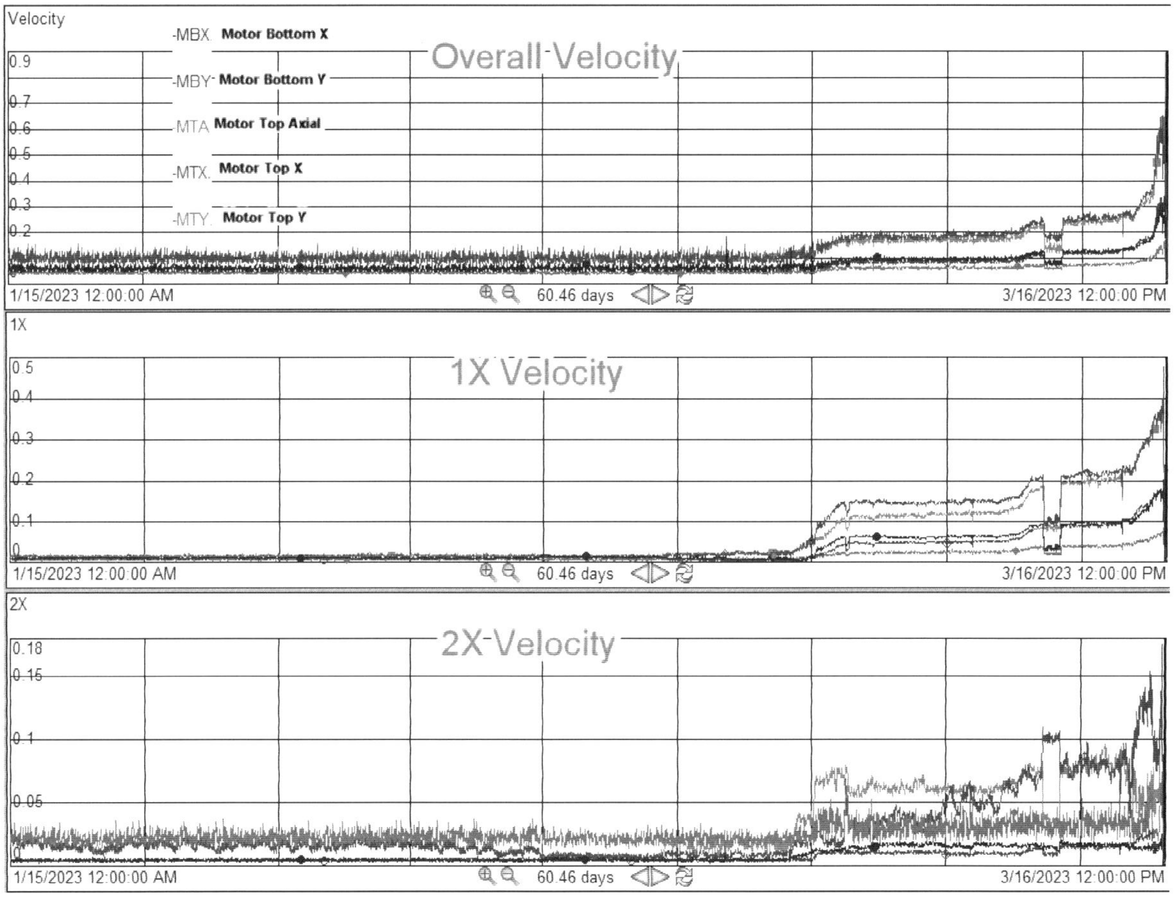

---FIGURE 16---

Driver: motor [1500 HP / 6600 V] upper bearings Kingsbury type thrust / sleeve guide; lower bearing anti-friction identified as 140BC02J3 – identified by site as 6228

Bore: 140 mm / OD: 250 mm roller diameter not specified

Frequency factors for NTN 6228 provided as: [0.418, 4.185, 5.815, 2.984] for [FTF, BPFO, BPFI, BSF]

Given identified FTF and 1X from data sets, the roller diameter that provides the best fit to these frequency factors is, d = 32 mm. This using the assumption that pitch diameter, D = average of bore and OD.

So, d/D = 0.1641.

For each data set, the analysis uses the two orthogonal lower motor bearing radial accelerometers, processing the spectral coherence using the analytic signal with X as real, and Y as imaginary. The traditional envelope just processes both channels and then for peak finding and parameter development a single spectrum from square root of sum of squares is used.

The d/D of 0.1641 is used and the contact angle that best aligns with the consistently present BPFI is used.

As the case was being tracked, less prescriptive approaches were followed in the event bearing factors for the specific version of a 6228, or the assumption that D = (OD + bore)/2 were not correct. However, confidence was gained, and the data sets were processed using the consistent approach described.

Plots

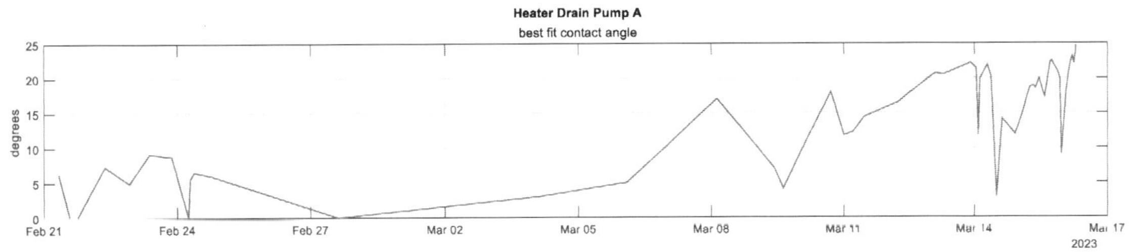

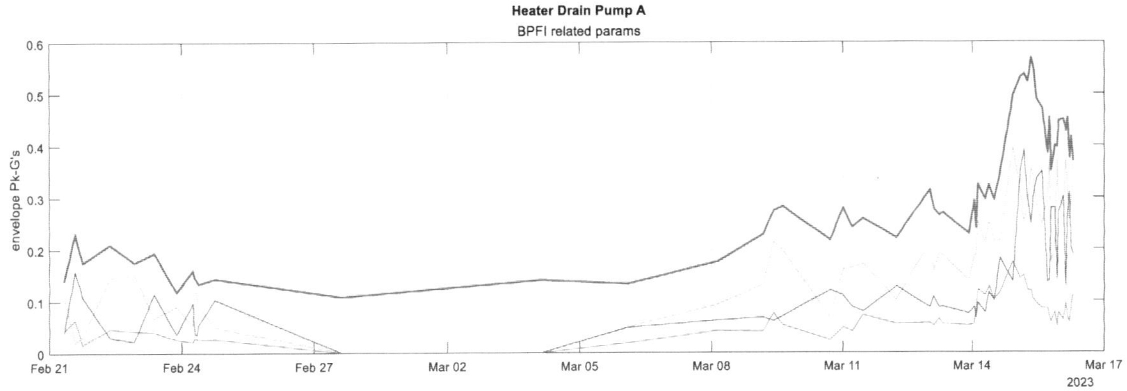

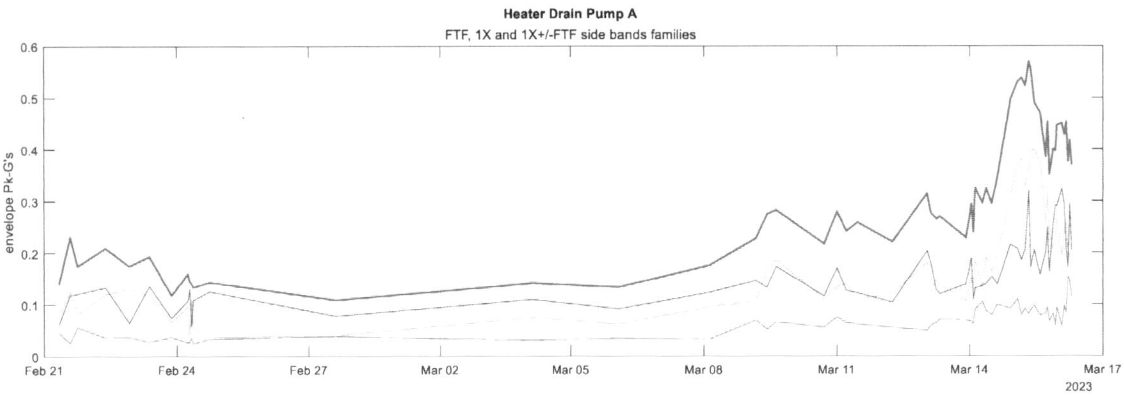

---FIGUREs 17---

BPFI-1Xsb (collection of matched peaks which are 1X side bands off of BPFI and 2xBPFI – the two carrier frequencies within the analysis band which was 0 – 500 hz). A consistent selection of the frequency band [2000 – 5800 hz] was used to avoid being swamped by a 120 hz result due to a strong carrier for 120 hz response in the 1500 hz range and Nyquist considerations associated with particulars associated with Spectral Coherence distribution using data sampled at 12800 hz.

The pump was removed from service due to exceeding termination criterion that used data collected with calibrated equipment by site personnel.

---FIGURE 18---

Lower Mtr bearing
Note that inner ring off center w.r.t. outer

Four balls were found broken in half

Case History (this motor is still in service and is used in walk through example above):

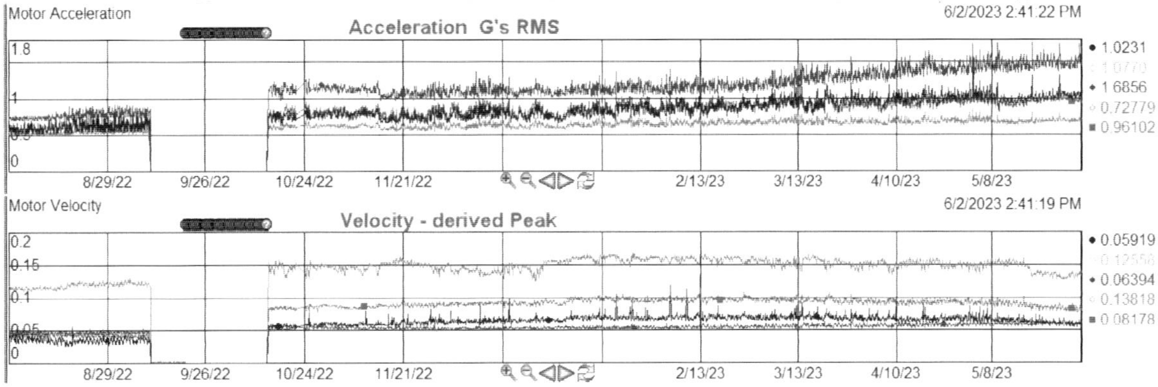

---FIGURE 19---

Noted the increase after a refueling outage and have been tracking this ever since:

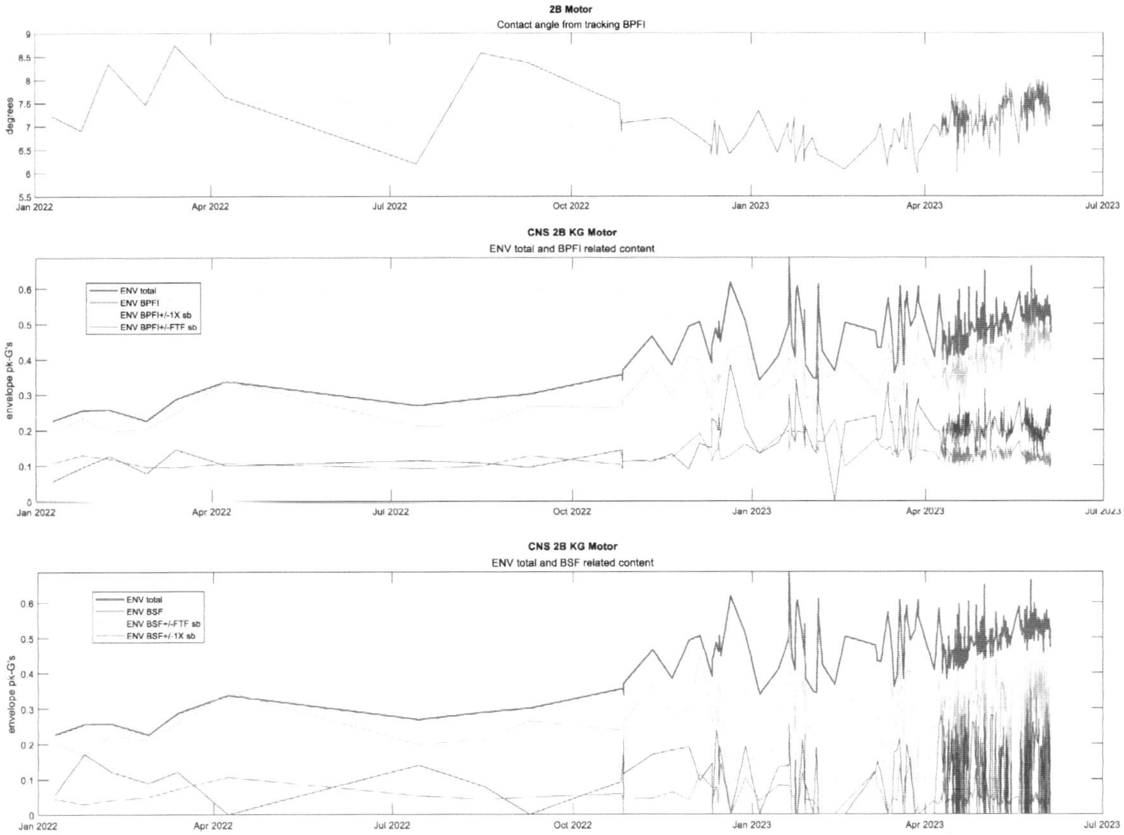

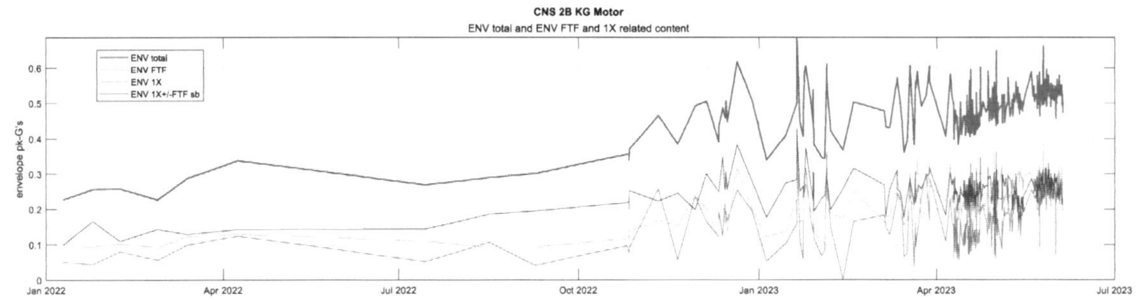

---FIGUREs 20---

Closing: It is not obvious that this effort is going to save lots of money. However, so far, when a bearing issue comes up in the fleet, the confidence installed by supporting a dialog that can demonstrate with clear plots and very accurate hits shown with clean cursors has been notable. When good identifications have occurred when we didn't know the bearing but were able to suggest that it had 18 rollers and it turned out it did and exhibited inner race and outer race damage the credibility of our team was bolstered.

From an different issue just this week:
Able to show 1X and very likely FTF and relevance with top plot – but not able to match BPFI. It was enough to satisfy the site to act. These plots were made using the app discussed in this paper.

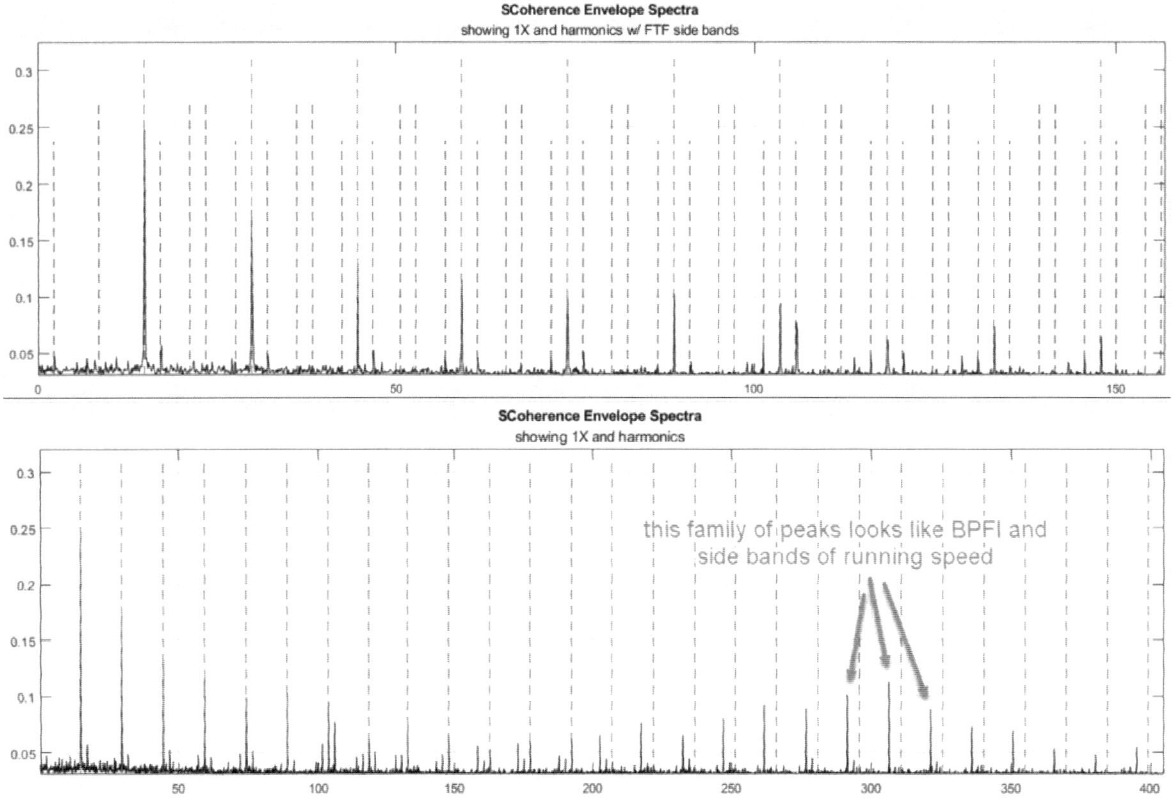

---FIGURE 21---

Change Detection for Improved Maintenance Notification

Eric Bechhoefer[1], Joelle Kessler[1]

[1] *GPMS International Inc, Cornwall, VT, 05753, USA*

Abstract: HUMS (Health and Usage Monitoring Systems) provide enhanced safety and availability by alerting maintenance personnel when components (shaft/gears/bearings) have degraded or become damaged. Proactive maintenance ensures that a degraded component will not fail in flight. Often, the damage is from high cycle fatigue or other relatively slow degradation processes. However, some events, such as maintenance, result in a step-change in the component's health. While high cycle fatigue degradation can be trended, a step-change in component health presents a challenge. This paper describes the development of two methods for step-change detection. These methods would allow the automated capture of maintenance or other events resulting in a step-change in component health.

Keywords: CUSUM, PELT, Health indicator, Maintenance alert, RUL, Smith Chart.

Introduction: Health and Usage Monitoring Systems (HUMS) provide benefits to both operational availability and safety, mostly in rotorcraft aviation. HUMS supports several functions that improve safety by proactively reporting the aircraft's performance. Those functions usually include notification of flight manual exceedances, flight data monitoring to support a safety management system, automated engine performance checks, drivetrain diagnostics, and rotor track and balance. These functions collectively allow HUMS to provide actionable information so maintainers can make data-informed decisions (Ref 1).

Mechanical diagnostics typically report on components that degrade due to fatigue damage. Fatigue damage can occur due to tensile cyclic loading. A power law can model the resulting fatigue crack growth (Ref 2). HUMS typically relies on signal processing techniques to generate component-specific condition indicators (CIs) (Ref 3). These CIs are mapped into a health indicator (HI). Trending facilitates both a threshold (e.g., alerting) and an estimate of the remaining useful life (RUL, Ref 4.). The HI and RUL are then used to provide alerts (e.g., actionable information) to the maintainer that prompt the planning of a maintenance activity to perform a corrective action. However, this trending and alerting capability do not address a step-change in component health.

A step-change in component health occurs when an event, such as maintenance, results in degradation or improvement to the HI value that is not associated with high cycle fatigue. That is, the HI trend and RUL modeled in Ref 4., do not adequately model events that are not associated with high cycle fatigue. This paper introduces a method of providing actionable information for drivetrain diagnostics due to a step change in component health.

This paper presents two examples of step changes in component health. In the first example, a Thompson coupling on the tail rotor driveshaft was assembled improperly. As a result of maintenance, there is a large increase in the 3rd harmonic of the shaft. This is indicative of a coupling problem. In the second example, an accessory drive duplex bearing degrades due to oil contamination. The performance of maintenance resulted in a step change event. In both cases, step change detection would have enhanced alerting and trending by alerting the maintainer as soon as possible that corrective action was needed.

It is proposed that a step-change is a case where the high cycle fatigue model is no longer in a state of "control." That is, the measured HI diverges from the filtered HI trend. Therefore, the high-cycle fatigue model no longer represents the damage process. This divergence is approached as a statistical process control problem, and powerful tools have been developed to detect this type of event. In selecting a technique, it is important that it has a short average run length (ARL), and the expected number of subgroups until a control chart first signals with a low computational load.

Two analysis methods were tested. The first solution tested is the cumulative sum (CUSUM Ref 5.). The second method is non-parametric changepoint detection, the Pruned Exact Linear Time (PELT) method (Ref 6).

The Trend Model: For trending, it was assumed that the rate of change in HI was bandwidth-limited, piecewise linear, and stationary. That is, the measurement noise, plant noise, and update rate are constant. As such, a forward/backward α-β tracker was used. This is a two-state filter of the HI, and its derivative.

Given the assumption of stationarity, the α-β tracker is treated as a steady-state Kalman filter. The filter coefficients for the α-β tracker (used for HI and dHI/dt) can be calculated as:

$$\lambda = \frac{\sigma_w dt^2}{\sigma_v}, \tag{1}$$

$$r = \frac{4+\lambda-\sqrt{8\lambda+\lambda^2}}{4}, \tag{2}$$

Where the process variance is σ_w^2, and plant noise variance is σ_v^2. The filter gains are

$$\alpha_1 = 1 - r^2, \tag{3}$$

$$\beta_1 = 2(2-\alpha) - 4\sqrt{1-\alpha}. \tag{4}$$

The filter is then:

```
For i = 1:n,
     fCI = fCI + dCI x dt;
     rk = CI(i) - fCI
     fCI = fCI + alpha1*rk;
     dCI = dCI + (beta1*rk)/dt;
```

CUSUM Approach: The CUSUM is a sequential method for detecting a change in mean values. The average run length (ARL) is often used as a measure of performance for detecting change. The ARL for the CUSUM is better than a Stewart control chart when the shift is less than two standard deviations. In this example case, the deviation is referenced to the mean HI from the trend, which models high cycle fatigue. In that way, changes due to a maintenance event, or catastrophic event, will deviate from the HI trend and will be detected by the CUSUM.

The CUSUM is a type of detection problem defined by:

- α: the PFA, e.g., stating a change occurred when in fact, it did not,

- β: the probability of not detecting a change when in fact, it did occur, and
- δ: the amount of shift in the process that one wishes to detect as a multiple of the process standard deviation.

Note that α and β are calculated in terms of one sequential test, where one monitor, S_m, until there is either an out-of-control signal or the signal returns to its nominal value, the HI trend.

These values are used to define the decision parameter needed for the test. Because the change detection will be used for both a deviation high (something failed?) and low (it was repaired), the following sequential values are calculated:

$$S_{hi}(i) = max(0, S_{hi}(i) + d_i - k) \tag{5}$$

and

$$S_{lo}(i) = max(0, S_{lo}(i) - k - d_i) \tag{6}$$

Where $S_{hi}(0)$ and $S_{lo}(0)$ are 0, and d_i is the difference between the trend and the unfiltered HI value at the time i. When either $S_{hi}(i)$ or $S_{lo}(i)$ exceeds a value h, the process detects a change, and a new trend is started. The values k and h, are defined as:

$$k = \frac{\delta\sigma_d}{2} \tag{7}$$

Where σ_d is the standard deviation of the difference between the trend and HI, and

$$h = 2k/\delta^2 ln\left(^{1-\beta}/_{\alpha}\right) \tag{8}$$

Thomas Coupling with CUSUM: An operator of a HUMS-equipped Bell 407GX performed standard maintenance on the oil cooler Thomas coupling. Approximately 30 hours after the maintenance was completed, the HUMS sent a warning alert (HI > 0.75). Viewing the CIs that were used in the HI (shaft order 1, shaft order 2, and shaft order 3), it was found that the shaft order 3 CI was in warning. Shaft order 3 is associated with a coupling assembly issue or coupling failure. An inspection of the coupling showed that it was assembled incorrectly. A repair was made, and the aircraft returned to service. The HI dropped to a nominal level. Figure 1 displays the current HI and trend during high shaft order 3 measurements and post-repair.

While the trend does respond to the step change, it is bandwidth limited: the step change takes place at acquisition 2112, and is not detected until 2754, or 642 acquisitions and 30 hours later. Ideally, the jump in the HI due to maintenance would have automatically alerted the maintainer. At this point, an inspection could be effectuated immediately. Similarly, once the repair was made, an automated alert could notify the maintainer that the repair was effective.

For demonstrating the CUSUM for change detection, a value δ = 2.5 (standard deviation of 2.5), β = 0.05 (probability of missed detection), and α = 0.01 (probability of false alarm) was set. The calculated σ_d = 0.11 gives a k of 0.13 and h of 0.19. The resulting change detection results are seen in Figure 2.

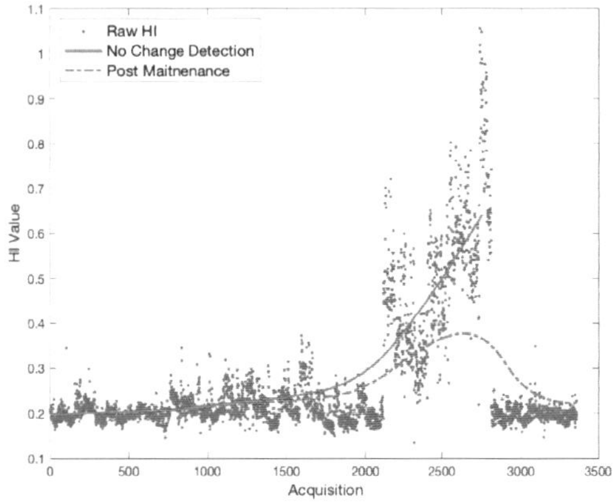

Figure 1 Damage Coupling No Change Detection

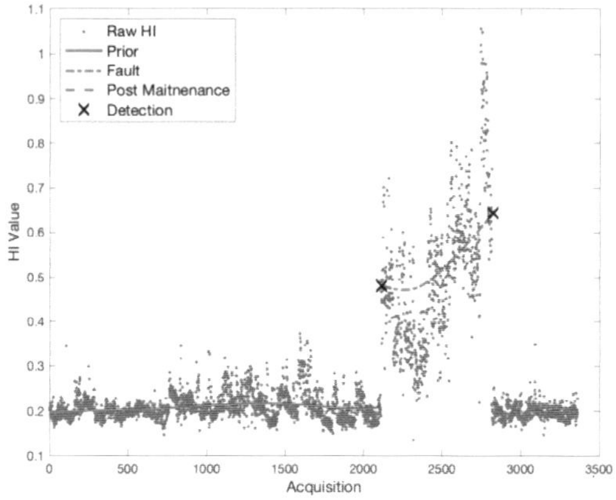

Figure 2 Coupling with Change Detection

The CUSUM change detection quickly alerts the change in HI due to maintenance almost immediately at 2113. Figure 2 captures these step-change events and would better alert the maintainer that the coupling requires additional maintenance or an inspection.

Aux Duplex Bearing with CUSUM: The second example is from a damaged auxiliary duplex bearing. The bearing is low in the gearbox and can pick up metal wear particles in the gearbox oil. It was hypothesized that wear particles contaminated the rolling elements. At approximately (x-axis) index 2425, the gearbox oil was replaced. The bearing rolling element energy (calculated using the envelope analysis, ref. 7) decreases from when the oil was replaced to acquisition 2900. Sadly, the damage was already done: the bearing fault propagates until acquisition 3312 (see figure 3)

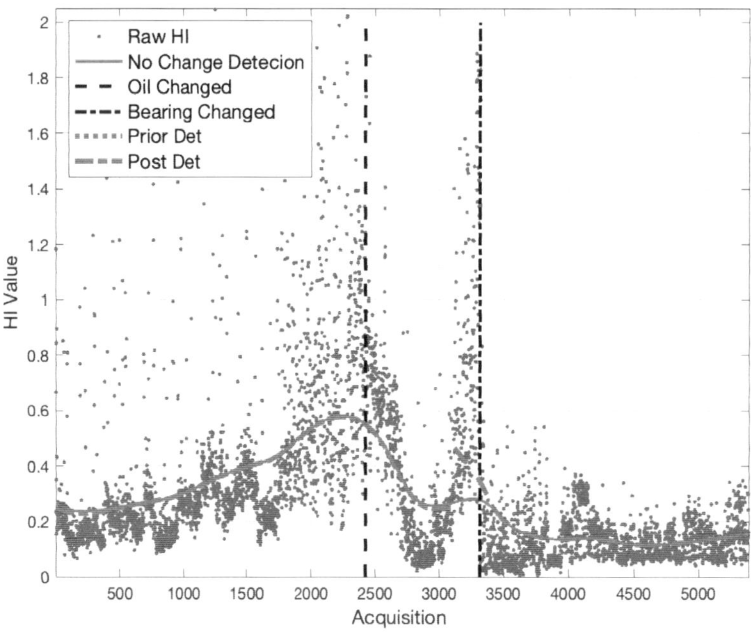

Figure 3 Aux Bearing Health with and without CUSUM Change Detection

In the Aux bearing example, the CUSUM failed to detect the step change due to the maintenance activities. This was likely due to the much larger σ_d calculated as 0.25. One assumption that is violated is that the standard deviation of the HI is not stationary. That is, the standard deviation is a function of time. This suggests that for a generalized change detection algorithm, a more robust calculation of σ_d (time-varying over some length of window) is needed or that techniques other than CUSUM may be more appropriate. For this reason, PELT was evaluated.

PELT for Change Detection: The Pruned Exact Linear Time method, as outlined in Ref 6, approaches change detection in a very different way than CUSUM. In PELT, the approach is to identify multiple change points by minimizing a cost function, *C*, for a segment of the data along a penalty function, *Bf(m)*, to guard against overfitting:

$$\sum_{i=1}^{m+1}[C(y_{ti-1}+1):ti]+Bf(m)\,] \tag{9}$$

Algorithmically, PELT inputs are:

A set of data, $y_1, y_2, \dots y_n$

A measure of the fit *C(.)*, which is dependent on the data,

A penalty constant *B*, which is not dependent on the number or location of the change points

And a constant *K*, which satisfies: $F(t) + C(y_{(\tau+1):s}) + K > F(s)$

The algorithm is initialized by:

n = length of the dataset,

$F(0) = -B$

$cp(0) = NULL$, and

$R_1 = \{0\}$

Interact for $\tau^* = 1, 2, \dots, n$

1. Calculate $F(\tau^*) = \min_{\tau^*} [F(\tau) + C(y_{(\tau+1):\tau^*}) + B]$
2. Let $\tau^1 = \arg\{\min_{\tau^*} [F(\tau) + C(y_{(\tau+1):\tau^*}) + B]$
3. Set $cp(\tau^*) = [cp(\tau^1), \tau^1]$
4. Set $R_{\tau^*+1} = \{ \tau \in R_{\tau^*} \cup \{\tau^*\}: F(\tau) + C(y_{(\tau+1):\tau^*}) + K < F(\tau^*)\}$

The output is the index of the change point.

Figures 4, 5, and 6 explore the change point detection of PELT at a different time in the fault progression. In Figure 4, PELT returns a change point at acquisition 2766. The change point is 2273, or about three hours before the maintenance event where the oil was replaced. If the trend is extended to aquation 3200, change detection how occurs at acquisition 2749. Note that the bearing appears to be propagating a fault at approximately acquisition 3000 (figure 5).

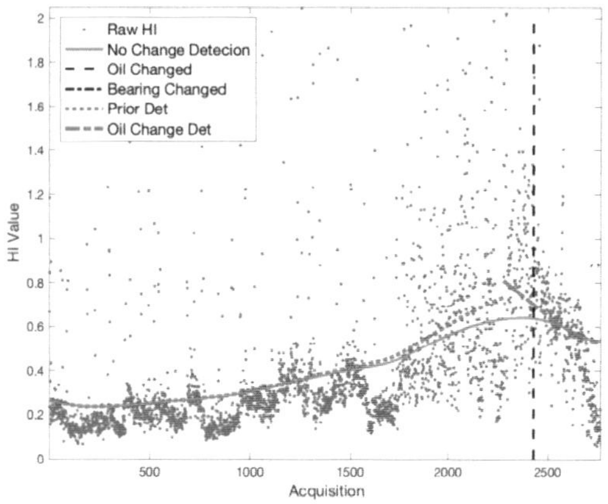

Figure 4 Change Detection After Oil Change

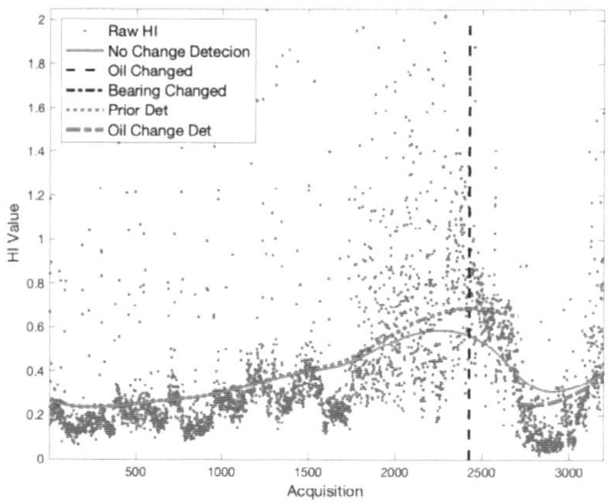

Figure 5 Change Detection Prior to Bearing Replacement

By acquisition 3390, or just 3 hours after the bearing is replaced, the change in the bearing HI is detected (figure 6). In general, PELT is detecting the changes in HI trend value, certainly better the CUSUM.

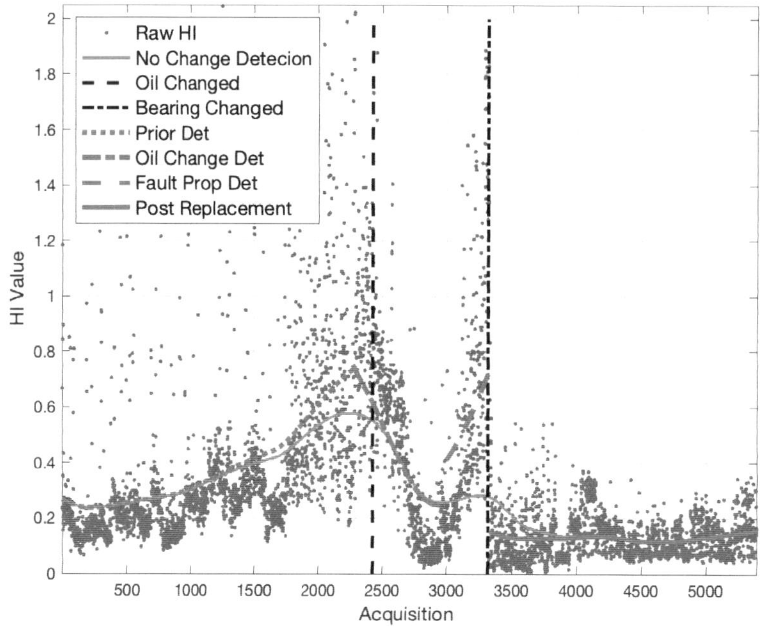

Figure 6 Change Detection Based on PELT

Discussion: Change point detection can provide actionable information to improve maintenance practice. However, this proves to be a challenging problem when trying to implement a method for all conditions. A major difficulty is that the variance is not stationary. Decision algorithms based on the changing mean value, such as CUSUM, are based on an estimation of the variance. In the Aux Duplex Bearing example, the variance is widely changing; as a result, the techniques failed to detect the maintenance actions related to the change of oil, the propagation of the bearing fault, or the step change when the bearing was replaced.

PELT implements a global optimization using dynamic programming. The cost function is not dependent on variance. As such, its performance is better. However, its run time performance is slower than CUSUM, which may pose an issue when doing analysis on large data sets (greater than 5000 to 7000 acquisitions).

Ideally, integration with a maintenance management system (MMS) could simplify the change detection problem. That is, the trend would be started when a particular component has undergone maintenance. For example, in the Thomas coupling example, that maintenance event should have started a new trend. In that case, there would be no need for change detection. Or, in the aux bearing example, when the oil was changed, the next acquisition should have started a new trend.

However, integration with MMS is a non-trivial task. For example, there would have to be a table for each maintenance action to link the action to a component in the HUMS database. This type of relationship is usually not present in the MMS. The MMS tracks hours and the required maintenance task to, at best, the line replaceable unit (LRU) but not to the component level inside the LRU.

As such, it depends on the HUMS vendor to improve the decision support tools by developing change detection algorithms.

References:

[1] HeliOffshore, "Health and Usage Monitoring systems HeliOffshore Recommended Practice Guidance," 2.0, 2020.
[2] Frost, N., March, K., and Pook, L., Metal Fatigue, Dover Publication, Mineola, NY, 1999.
[3] Večeř, P., Kreidl, M., Šmíd, R. "Condition Indicators for Gearbox Condition Monitoring Systems," Acta Polytechinca Vol. 45, 6/2005
[4] Bechhoefer, E., Xiao, L. Zhang, X., "Remaining Useful Life Calculation of a Component Using Hybrid Fatigue Crack Model", Proceeding of the Annual Conference of the PHM Society, Vol 12 No.1, 2021
[5] NIST/SEMATECH e-Handbook of Statistical Methods, 6.3.2.3, CUSUM Control Charts, https://www.itl.nist.gov/div898/handbook/pmc/section3/pmc323.htm.
[6] Haynes, K/, Fearnhead, P., Eckley, I., "A computationally efficient nonparametric approach for changepoint detection", Stat Comput (2017) 27:1293-1305.
[7] Randall, R., Antoni, J., "Rolling Element Bearing diagnostics – A Tutorial", Mechanical Systems and Signal Processing, Volume 25, Issue 2, p. 485-520, 2011.

MFPT ANNUAL CONFERENCE

Where Theory meets Practice

IDENTIFYING MACHINERY ANOMANLIES USING SHAPE IDENTIFICATION AND CLASSIFICATION ALGORITHM

Mantosh Bhattacharya
Petrofac International UAE
Mantosh.b@petrofac.com

Mantosh Bhattacharya is designated Subject Matter Expert on Turbomachinery in Petrofac Intl. located in UAE. He is a chartered engineer CEng (MIMechE-UK), Intl. PE (UK), a member of ASME, Vibration Institute and ISCM – (BINDT UK). He is a member of the board of directors of MFPT (Society for Machinery Failure Prevention Technology) USA, advisory committee of ATPS [TEES – Texas A&M] and HI 9.6.4, 9.6.5 and 14.5.1 standards. He holds 3 U.S. patents related to structural dynamics. He has published and presented several technical papers in various journals and conferences.

ABSTRACT:

The prime responsibility of anomaly detection and mitigation of rotating machinery distress lies with a certified vibration analyst. Vibration analysis is a multi-layered task and with huge number of assets in a large plant, the work can become tedious. With the introduction of a machine learning algorithm, the first layer of the task can be offloaded to an AI based module. The direct unfiltered numerical values of vibration with units such as mm/sec RMS, ips RMS, micron pk-pk, mils pk-pk which is normally used as marker for alarm and trip values. However, for anomaly detection, vibration signatures are analyzed in forms of various plots like Lissajous (Orbit) plots, band based FFT spectrums and shaft centerline plots. This paper presents a road map for detection of anomalies using a simplified multi-layered neural network technique based on geometrical functions of vibration signature plots. From a continuous streaming data, the algorithm can detect the plots of concern from a dataset of synthetic images designed to benchmark systems for understanding the spatial and logical relations among multiple shapes.

This paper deals only on the cyclo-stationary-periodic signal for maintaining simplicity in understanding for readers.

KEY WORDS - Spectral Band alarms; Orbit plots; SVM ; Convolutional neural network

INTRODUCTION TO DAMAGE INDICATORS – DATA & VISUAL

Figure 1 (p-f curve) shows a prognosis of damage of rotating machinery where due to many external reasons the machinery distress starts. The P-F curve displays the health of equipment over time to identify the interval between potential failure and functional failure. Point one to point three is the range of predictive maintenance where three stages of potential failure symptoms are identified. This slope can be flatter (increase in life span of machinery) based on actions taken by the reliability centered maintenance team at those stages (supported by experienced vibration analysts). Conversely, if proper maintenance is not undertaken or support of analysts is not provided, the curve gets steeper, and machinery plunges into a catastrophic breakdown rendering high financial impact due to loss of asset.

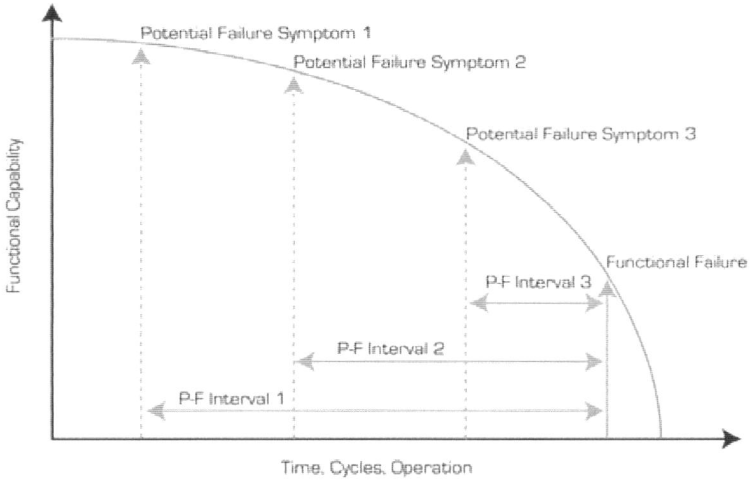

Fig 1 - A graphical representation of multiple P-F intervals. [Courtesy Reliable plant]

Figure 2 highlights various standards from which data-based damage indicators can be derived such as ISO 10816 and ISO 7919-3

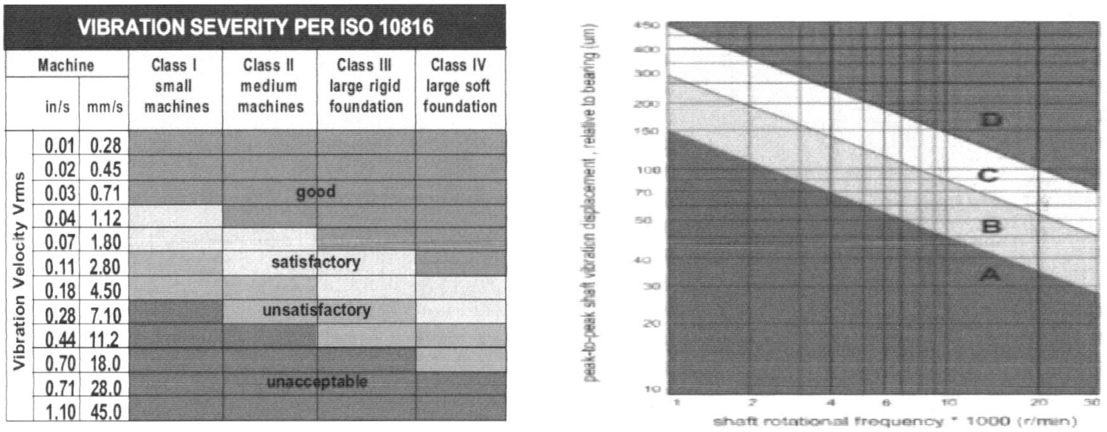

Fig 2 – Machinery vibration severity charts base on ISO 10816-3 and ISO 7919-3

Alarm and trip settings are normally based on the above ISO standards or by OEM based on their database of machinery health matrix. However, a robust condition monitoring team embraces pre-alarm philosophy where operators are alerted and the diagnostic team starts working to find the root cause and take mitigation action to avoid early breakdowns.

Numeral based damage indicators alone cannot provide enough insight of the sustained damage which allows the analyst to predict root cause and label the issue based on severity so that the operation team takes proactive steps. Visual based damage indicators considered in regular analysis are FFT spectrum, water fall plots, shaft centerline plot, polar plots and Lissajous plots (shaft orbit plots) as shown in figure 3which are displayed in diagnostic platform of machinery vibration monitoring system.

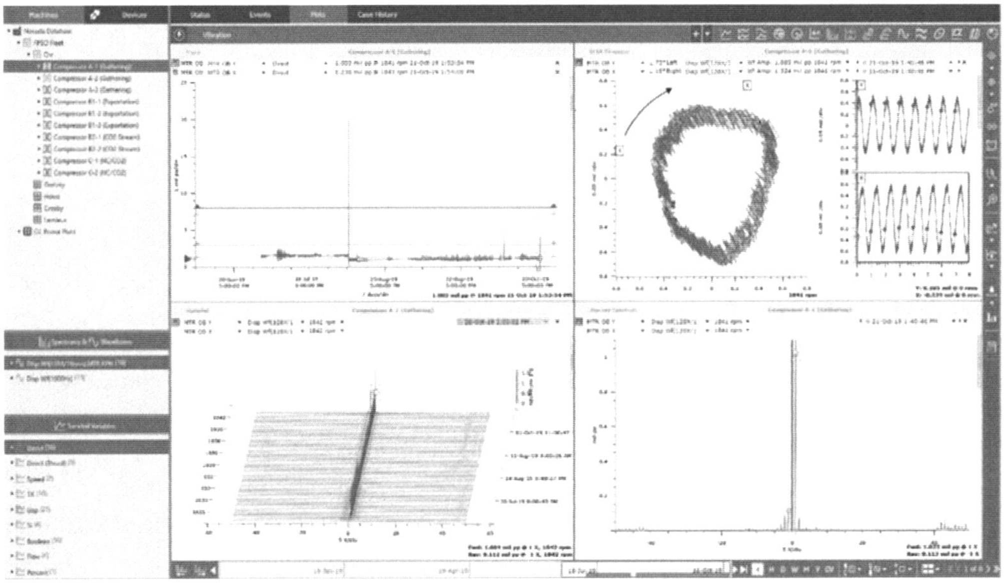

Fig 3 – Visual based damage indicators as seen on diagnostic platform

HOW ARTIFICIAL INTELLIGENCE (AI) CAN HELP

Given a vibration signal, an engineer may be able to do some exploratory analysis but this requires a manual analysis of the signal in various forms. Sensors measure a broad range of parameters, such as pressure, flow and temperature and generate very large streaming data every day, however the number of experts to decipher the sensible data are not readily available. Rotating machines form a vicious fault cycle which is why it is essential that the faults are recognized in their formative stages, so that remedial measures can be initiated on time.

A digital assistant similar to a certified CAT II vibration analyst can help the vibration analyst during their day-to-day work. This is where AI-based methods can make reliable first and second level decisions without the presence of a machine diagnosis expert. Several AI-based methods of vibration analysis for machine monitoring and diagnosis are such as support vector machine (SVM) and neural network (NN) fuzzy logic, to name a few. It is important to quantify the process and vibration signals into a handful of metrics, then track and compare them.

These metrics should consider primary system aspects like:

- Accessing crucial process data
- Accounting for process variations
- Correlation with shared process variations and vibration data

The first part of the proposal is to have a trigger-based data acquisition system to avoid capturing monotonous data. The second part is to utilize spectral band alarms to analyze the sensible data from spectral content and analyze associated shaft orbit plots based on various machine learning algorithms. The third part describes the plausible shape identification and classification algorithms which can work as AI based assistants for SMEs (subject matter expert) to reduce workload.

SIGNAL ACQUISTION – TRIGGER BASED DATA CAPTURE

A pre-alarm value of overall vibration can be set as a trigger in the system and when this value is crossed, the system will begin to capture and stream the data. The system will start accounting for process variations to help the tool corelate the values for proper diagnostics. Few crucial process variables can be used as weightage factor when we move to AI based identification and classification for diagnostics. A baseline equipment response data set with terms of numerical value and plot is a useful tool which can help establish alarm levels and focus on probabilistic faults.

Band alarms

Certain vibration frequencies (relative to the machine rotating speed) relate to specific group of machine faults such as rotor mass imbalance, misalignment like 1x, 2x and 3x rpm peaks. These type of vibration peaks fall in a certain band of frequencies. Other frequency bands associated with various faults typical of some specific machine can also be targeted to trigger alarms. If there are changes in machine speed, then orders are used and spectral bands can be generated. In this concept any vibration crossing a limit assigned to a particular band can trigger a warning (alert) and alarm. This method can provide the vibration analyst / software time to process and classify faults.

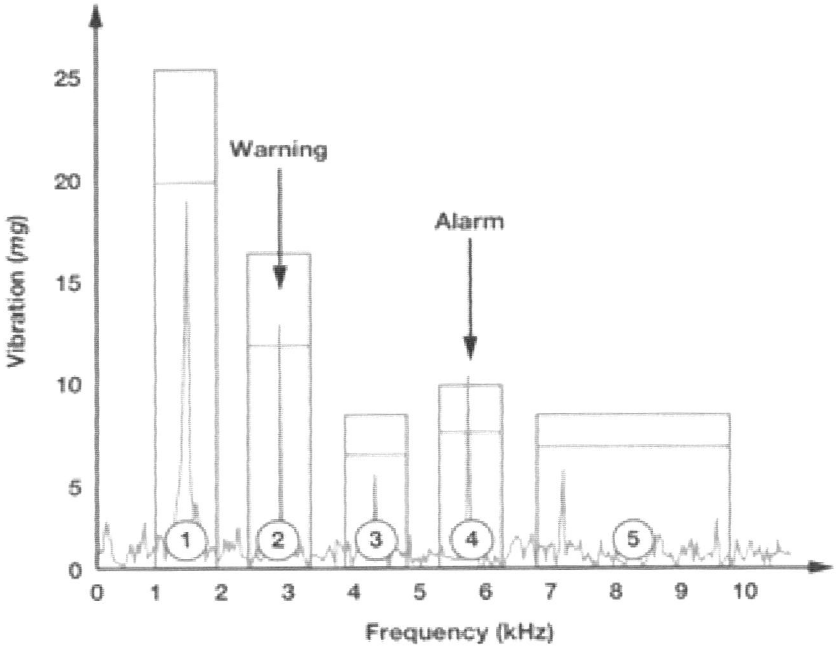

Fig 4 – Band based spectral alarm

ORBIT PLOTS

An orbit plot is the visual representation of the centerline of a rotating shaft. The shape of the orbit often holds important clues as to the source of excessive vibration. An incipient malfunction of the machine generates the unique orbit pattern. Therefore, an understanding of orbit shapes can help identify how the dynamics of machinery malfunctions can take place and how they can be more accurately detected before failure.

Although the orbit analysis (orbit shape and size) is commonly used to diagnose rotating machinery, the diagnosis heavily depends on the knowledge of an expert for classification and shape identification to detect incipient fault in the rotating machine.

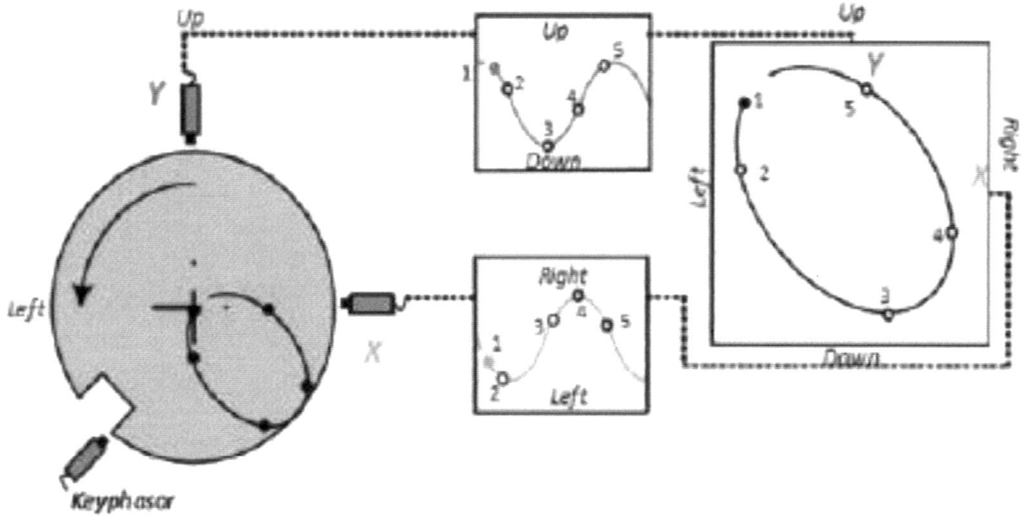

Fig 5 – Synthesis of a shaft Orbit plot

Geometrical features of an Orbit plot

Normally 1x filtered orbits are an elliptical shape as shown in Figure 6. The ellipse has two focal points, also called foci. The eccentricity of the ellipse lies between 0 to 1. The total sum of each distance from the locus of an ellipse to the two focal points is constant. Ellipse has one major axis and one minor axis formed by two sinusoidal timewaves. The ratio between minor and major axis play a major role in classifications of various elliptical shape. Orbits are also classified / recognized by SDI (Shape and Directivity Index) in a machinery train.

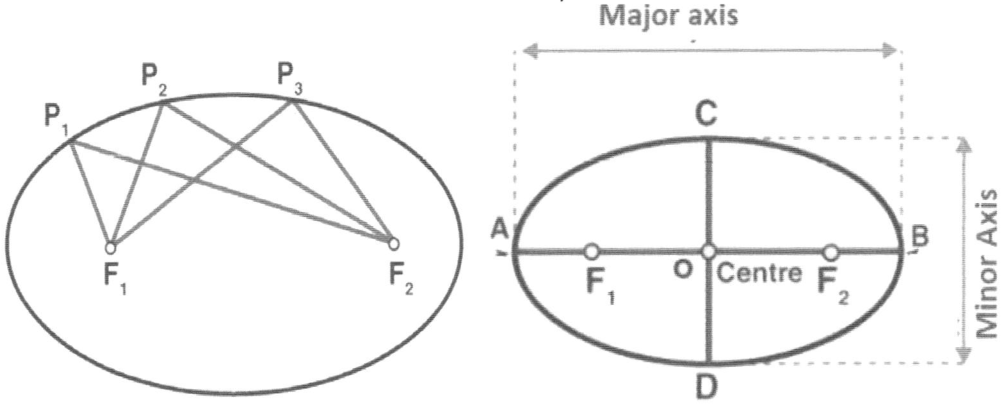

Fig 6 Geometrical features of ellipse Photo courtesy – Byjus's math class

The ellipticity of the filtered orbit, as well as the amplitude and the inclination angle of the major axis and minor axis of the orbit are parameters that can provide important diagnostic information. The known method to estimate these parameters is the standard Hough transform. The Hough transform is a feature extraction technique used in image analysis, computer vision, and digital image processing. The technique identifies imperfect instances of objects within a certain class of shapes by a voting procedure.

AI BASED METHODS

In this paper, we propose a supervised machine learning method to autonomously identify different orbit shapes generated by rotating machinery so that a more robust and automatic monitoring system can be established. The identification of shaft orbit can be treated as a pattern recognition problem of two-dimensional (2D) images.

Supervised learning "learns" by adapting itself to minimize errors. After training, the performance of the learned relationship (function) are evaluated on a test set (of examples) that is separate from the training set. Supervised learning is useful for pattern recognition, function regression, etc. One example of applications is recognition of handwritten numbers which are 2D plots. One of the major complex tasks it performs is identifying the handwritten numbers 4 and 9 which are of very similar feature. A supervised classifier is trained with a huge data base of handwritten numbers, each with a label (the true number each image represents). Once validated on a separate test set, the trained classifier is used for fast and accurate recognition of handwritten numbers, and hence this becomes a strong candidate as a suitable algorithm for orbit shape recognition and classification.

The paper has focused on the most popular methods such as Support Vector Machine (SVM) and Deep Learning using convolution neural network. The selected methods are chosen based on their efficiency in face recognition, handwritten digit identification and classification. The methods utilize salient geometrical features of shaft orbit plots as a 2D image. A ML (machine learning) engineer was consulted for writing codes to run the algorithms which is proposed in this paper. To avoid commercialism in this technical paper, algorithms incorporated using few software are not mentioned explicitly.

Dataset of good and bad orbit plots for fault diagnosis using AI

The shaft orbit of different shapes reflects the operation of different units, and its shape is an important basis for judging the running state and information of the unit. Figure 7 shows the direct orbit shaft orbit shapes and Table 1 shows the potential failure types the digital assistance can work as advisory.

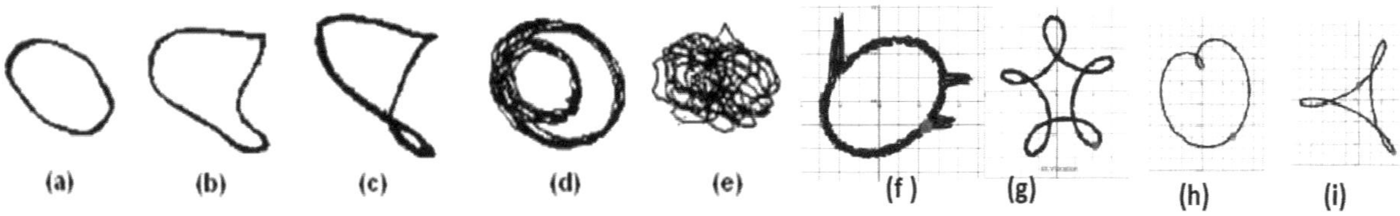

Fig 7 – Shapes of direct and Filtered shaft orbit plots and classification.

The identification of shaft orbit can be treated as a problem of the pattern recognition of 2D images for fault diagnosis.

Table 1 - Shaft orbit shapes (Direct and 1x compensated) and potential failure types.

SN	Orbit plot shape	Failure types based on expert knowledge
1	Constricted Ellipse (1x filtered)	High unbalance, high asymmetric stiffness caused by external issues
2	Ellipse (1x Filtered)	Unbalance, standard asymmetric stiffness
3	Circular shape with large diameter (> 75% of direct value)	High unbalance or passing at resonant speed
4	Sharp spikes (unfiltered)	Dent on probe tracking area
5	Orbit plots with more than one dot (key phasor marker)	Oil whirl. Many dots indicate shape of chaos", mostly corresponds to the oil whip.

6	Banana (Unfiltered) / Flatness in Orbit – Truncation	Resultant fault of severe rotor unbalance and severe shaft misalignment, rubs
7	Inner 8 (1x filtered)	Commonly caused by oil-film whirl
8	Outer "8" (filtered)	Shaft misalignment
9	Petal shape	Fractional whirl (sub synchronous vibration)

DATA SETS

The detection of ellipses in digital image data is an important task in vision-based systems established on the above-mentioned damage mechanisms. The dataset to be used in this proposal can be generated from bench marks, case studies or machinery fault simulator (MFS).

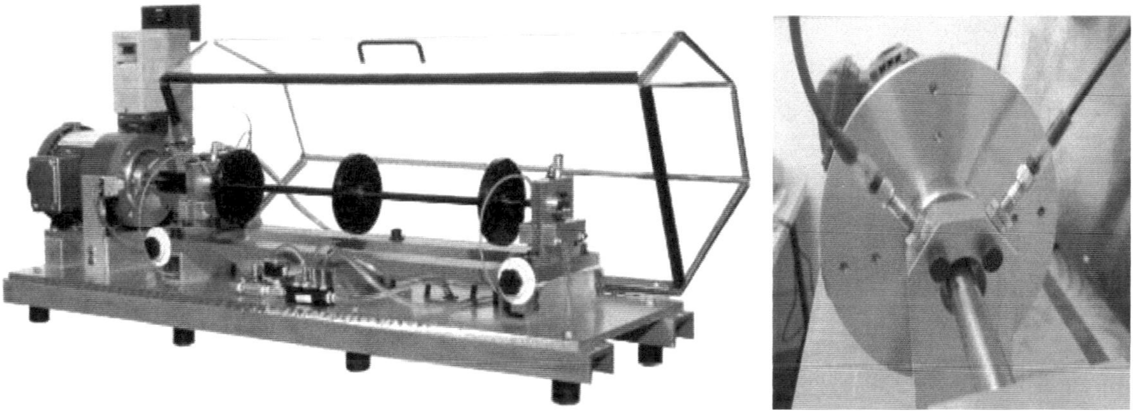

Fig8 – Machinery fault simulator (left) X-Ysensors (right) courtesy -Spectra quest Inc.

Figure 9 shows various orbit shapes which are considered as damage indicator or precursor machinery distress the dataset of orbit comprise of images of orbit which are categorized into circles, ellipses, banana shapes, truncated shapes, 8 shape, petal like shape and circle inside circle.

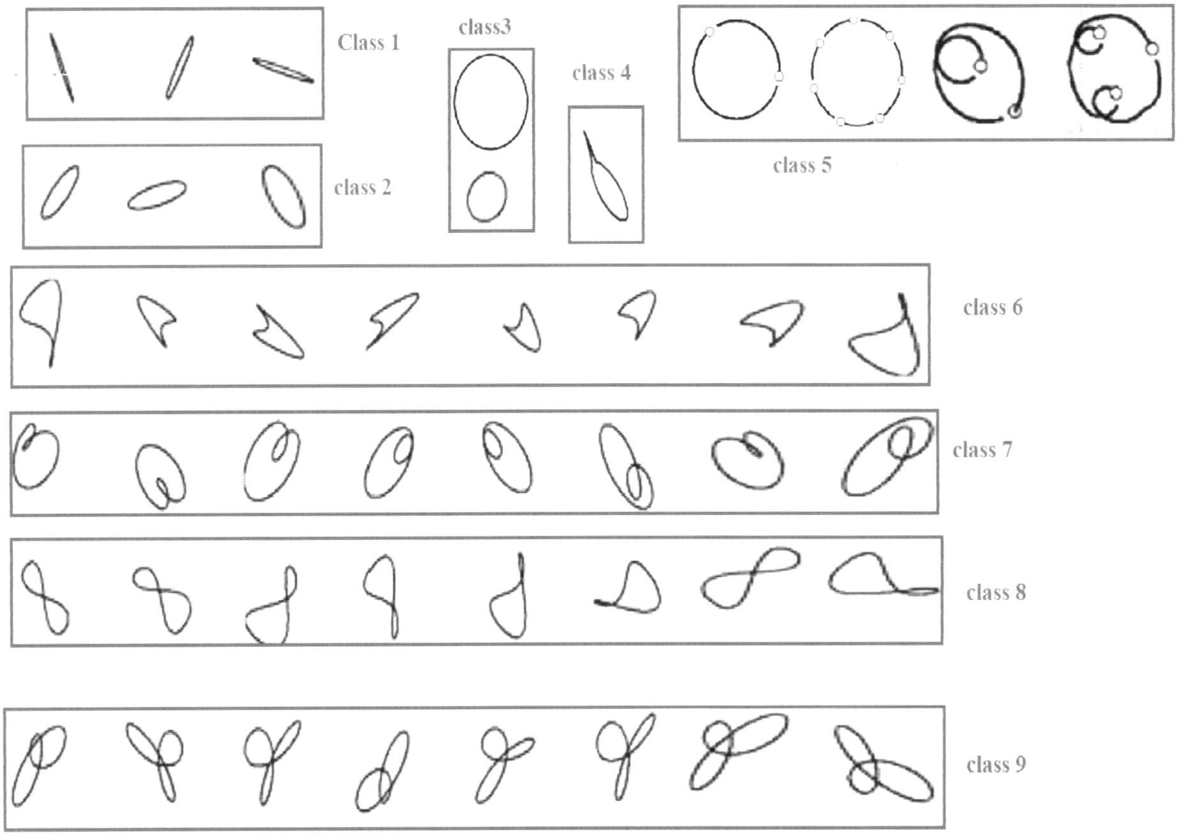

Fig 9 - Shape based data base of various types of orbit shapes as input for training set

First, the region boundary is smoothed by applying a Gaussian filter, a low pass filter used for reducing blurring regions of an image.

The shapes can be defined based on coordinate systems using 9 classifiers and creating an SVM to detect the shape of plot.

Method - We now consider how the orbit shape classification algorithm can help identify probable damage index using Support Vector Machine (SVM).

SVMs are a powerful machine learning algorithm that can be used for classification tasks. They work by finding the hyperplane that maximally separates the different classes of data in a high-dimensional feature space. SVMs have been successfully used in many applications, including image classification, text classification, and fault diagnosis of rotating machinery. Considering the task of identification and classification of various type of Orbit plots are similar to task of face recognition, the suitable opensource programming language can be used.

The coding process for using the SVM involves importing the Orbit plot dataset so that most probable fault can be listed. This dataset should consist of around 4500 labeled images with each having a label associated with it. Each category consists of 900 images. The training set of each category contains 700 images and the testing set contains 200 images. This dataset has 4500 images each with 1024 features since each image is considered to be 32x32 pixels and each feature represents one pixel's intensity. These 2D images will be various orbit plots.

The steps for the coding process are as follows:

First, import the dataset, plot some of the orbits to understand the dataset, and split the dataset.

Now comes the important part, we will be using the Support Vector Classifier (SVC) in order to classify these plots into their corresponding categories. Here, the data will consist of the matrix representation of all the plots of the dataset which can be understood by the SVC algorithm. Now, we can clean dataset to remove any unwanted information and noise. Then, we train and test the SVM to check accuracy score and to find if there are still possibilities to improve the accuracy of the model.

By looking at the classification report we can figure out the precision using the confusion matrix with four quadrants defined as True Negative (TN), True Positive (TP), False Positive (FP), False Negative (FN).

It is important to experiment with different kernel functions and hyperparameter values to find the best combination that maximizes the model's accuracy.

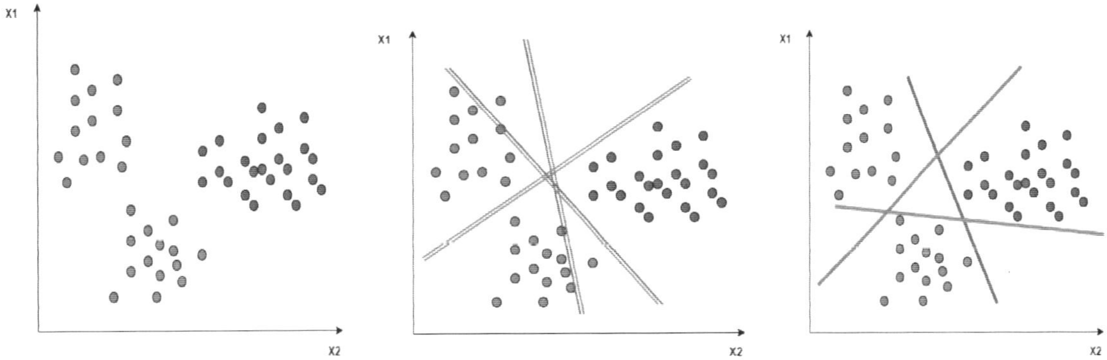

Fig 10– Multiclass classification approach by SVM

In multiclass SVM, the algorithm is trained on a dataset that has multiple classes. The goal is to classify new data into one of these classes. One approach to solving this problem is to use a one-vs-all (OVA) approach, where a binary SVM is trained for each class. This means that for each class, the SVM is trained on all the data that belongs to that class, as well as a subset of the data from the other classes.

ARTIFICIAL NEURAL NETWORK (ANN)

The ANN of orbit shape recognition can be trained by the training patterns generated by computer simulation for plenty of orbit shapes. The ANN is a well-established classification method that can be applied for fault detection and machine diagnosis in rotating machines.

A typical neural network with hidden layers is shown in fig 11.

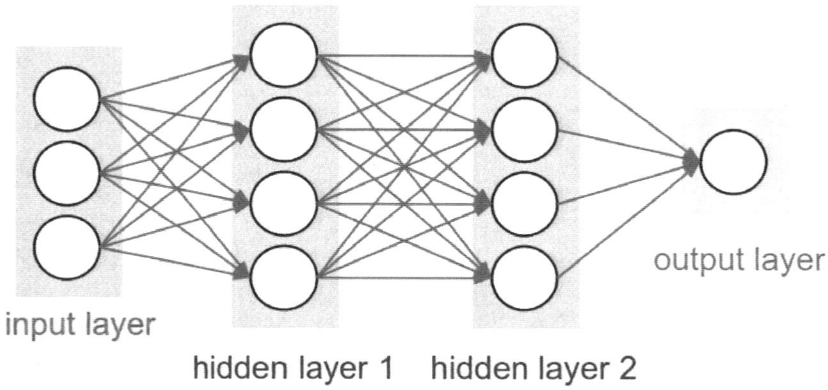

Fig 11 – Sample feed forward neural network with hidden layers.

A neural network is a set of connected input/output units in which each connection has a weight associated with it. During the learning phase, the network learns by adjusting the weights in hidden layers to be able to predict the correct class label of the training input.

A feed forward back-propagation algorithm is the standard algorithm for any supervised-learning pattern recognition process and is based on the weight error correction rules. The autonomous pattern recognition and classification algorithm may use the deep learning method on shaft orbit shape images.

Adding weightage to a hidden layer in a neural network involves adjusting the connection weights between the input layer and the hidden layer. The weightage determines the strength of the connection between each input and the hidden layer neuron, which affects how much each input contributes to the overall output of the hidden layer.

The weights can be initialized randomly and then adjusted during the training process using a technique such as backpropagation. During backpropagation, the network is trained using a set of labeled input-output pairs, and the weights are adjusted iteratively to minimize the difference between the predicted output and the true output for each input.

A convolutional neural network is implemented to construct weights between neurons and to generate the entire structure of the neural network. Then, the created network enables us to classify fault modes of rotating machinery via orbit images such as raw, slow roll compensated orbit and 1x and 2x filtered orbit shapes. Fig 12 shows how handwritten digits are identified with neural network.

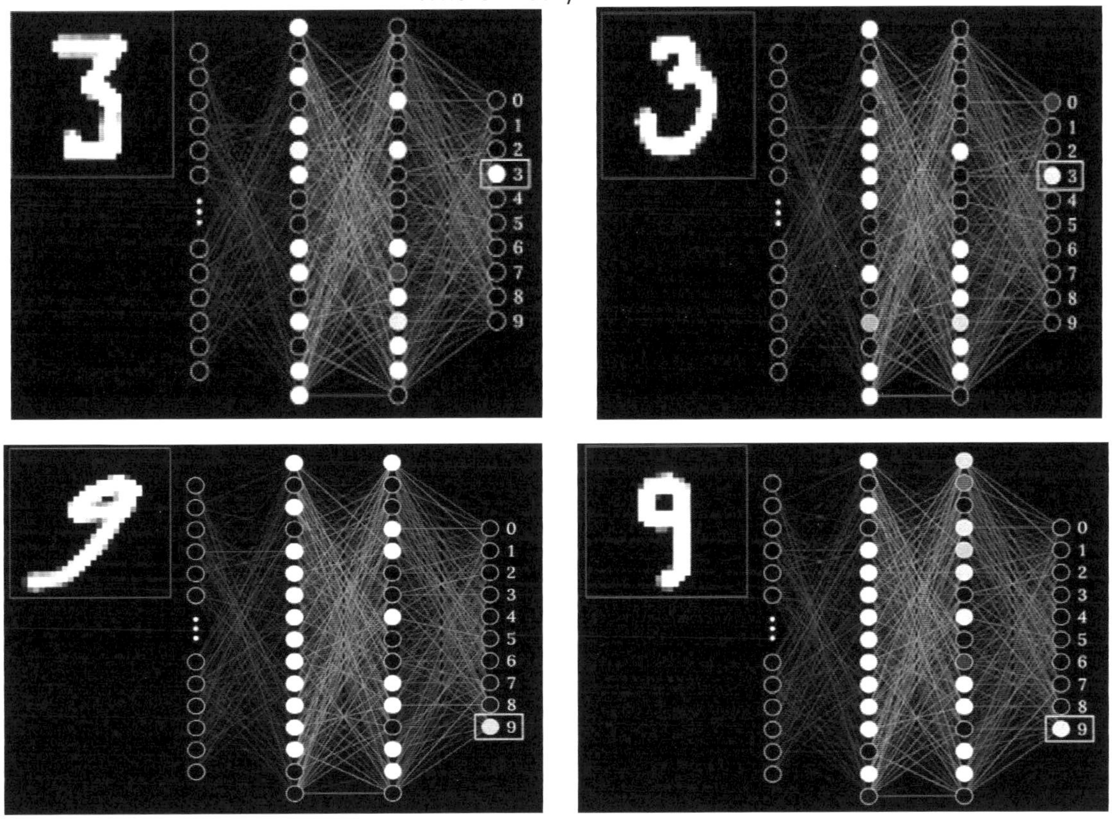

Fig 12 – Handwritten digit identification using neural network courtesy 3blue1brown

A convolutional neural network (CNN) is a type of Artificial Neural Network (ANN) used in image recognition and processing which is specially designed for processing data(pixels). This helps to recognize the image which has complex problems for edge detection.

An intelligent diagnosis method combines feature selection and the DNN (Deep neural network) method. The feature selection method is based on the distance between the features to select the sensitive features. Its selection principle is that the distance between the same classes is the smallest and the distance between the different classes is the largest.

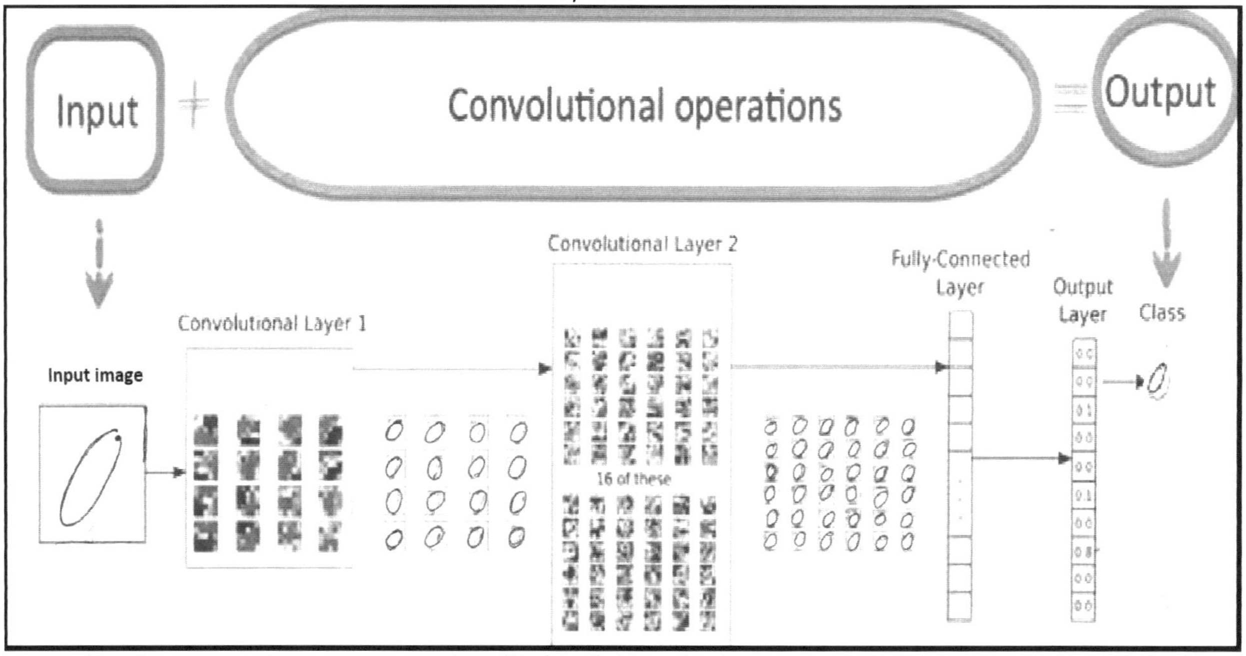

Fig 13 – Convolution based neural network for detection/classification.

OBJECT RECOGNITION

An algorithm to determine what object classes are present in the image is required.

- Image classification: Predict the type or class of an object in an image
- Input: An image with a single object, such as an easily recognizable plot with defined geometric features.
- Output: A class label (e.g., one or more integers that are mapped to class labels)
- Image classification: Algorithms produce a list of object categories present in the image

Once the image of shaft orbit is identified as per certain class, Thus, a trained neural network may be thought of as a virtual "CAT II vibration analyst" for the category of information it has been given to analyze.

CONCLUSIONS & IMPLEMENTATION

Currently, it is the responsibility of the subject matter expert to find the best algorithm for the respective application with minimal implementation effort for various applications for machine condition monitoring. An important consideration is the integrity of the sensor data itself. Without a reliable sensor, identified performance shifts could be attributed to either the equipment or the sensor.

It is recommended to implement a selected machine learning by starting with just one asset or a particular class of assets as a pilot project with a scope to adjust weightage or back propagation neural network. A properly scoped project allows you to determine if the project is yielding the intended results.

To set up the system appropriately, we should look at the predictive analysis and verify the prescriptive actions with the designated team. Experts in the field will leverage their knowledge for creating weightage and biases for neurons. As we grow with back propagation and proper weightage the model will evolve to higher maturity.

AKNOWLEDGEMENTS:

The author wishes to tender sincere thanks to Mr. Arnav Bhattacharya (Masters in data science and expert in AI based projects), who had taken up to run many algorithms to demonstrate the feasibility of proposal. Additionally, author thanks Mr. MVL Narasimham and Mr. Vineet Khanna of Petrofac for internal review and suggestions for structuring of this paper. The author also acknowledges the support of MFPT 2023 review committee for accepting this paper for MFPT 2023 conference.

REFERENCES

1. Intelligent fault diagnosis of rotating machine elements using machine learning through optimal features extraction and selection -Syed Muhammad Tayyaba , Eram Asgharb, Paolo Pennacchi c, Steven Chatterton -30th International Conference on Flexible Automation and Intelligent Manufacturing (FAIM2021) 15-18 June 2021, Athens, Greece.
2. Basic Machine Learning course notes – TATA Steel India
3. Ellipse Detection in Digital Image Data Using Geometric Features - Lars Libuda, Ingo Grothues, and Karl-Friedrich Kraiss
4. Multi-Fault Diagnosis Of Industrial Rotating Machines Using Data-Driven Approach: A Review Of Two Decades Of Research- Shreyas Gawde,Shruti Patil, Satish Kumar et al
5. Vibration Analysis for Machine Monitoring and Diagnosis: A Systematic Review - Mohamad Hazwan Mohd Ghazaliand Wan Rahiman
6. Intelligent fault diagnosis of rotating machinery based on deep learning with feature selection -Dongying Han, Kai Liang and Peiming Shi https://doi.org/10.1177/1461348419849279
7. A Gentle Introduction to Object Recognition With Deep Learning by Jason Brownlee
8. A review of vibration machine diagnostics by using artificial intelligence methods - Grover V. Zurita and Diego Román Cabrera https://www.researchgate.net/publication/314782455
9. Enhanced K-Nearest Neighbor for Intelligent Fault Diagnosis of Rotating Machinery - Jintao Lu, Weiwei Qian, Shunning Li and Rongqing Cui
10. Bently, D. E., Hatch, C. T., “Fundamentals of Rotating Machinery Diagnostics”
11. Image Categorization by Learning and Reasoning with Regions-Yixin Chen and James Z. Wang

12. Rotating Machinery Diagnostics using Deep Learning on Orbit Plot Images - Haedong Jeong, Seungtae Park, Sunhee Woo, and Seungchul Lee 44th Proceedings of the North American Manufacturing Research Institution of SME http://www.sme.org/namrc
13. Diagnostic Significance of Orbit Shape Analysis and its Application to Improve Machine Fault Detection - N. Bachschmid, P. Pennacchi and A. Vania
14. Human Face Shape Classification with Machine Learning- Ashwinee Mehta and Taha Mahmoud https://www.researchgate.net/publication/362903132
15. Shaft orbit pattern recognition for fault diagnosis of Induction motor using Neural Network.
16. Conference Paper · December 2011-Pratyay Konar and Paramita Chattopadhyay https://www.researchgate.net/publication/269280012
17. Automatic Detection and Recognize Different Shapes in an Image -Nidhal El Abbadi and Lamis Al Saadi
18. Real-Time Ellipse Detection for Robotics Applications - Azarakhsh Keipour, Guilherme A. S. Pereira, and Sebastian Scherer
19. Machine Fault Diagnosis and Prognosis: The State of The Art -Tran Van Tung and Bo-Suk Yang International Journal of Fluid Machinery and Systems -Vol. 2, No. 1, January-March 2009
20. Geometrical and Graphical Representations Analysis of Lissajous Figures in Rotor Dynamic System - Hisham, Al Kahazali and Md. Askari.
21. Recognizing Geometric Shapes in 2D Space Using Machine Learning
22. But what is a Neural Network? Lesson by Grant Sanderson
23. Recent advances in the application of deep learning for fault diagnosis of rotating machinery using vibration signals- Bayu Adhi Tamal,Malinda Vania,Seungchul Lee· Sunghoon Lim, https://link.springer.com/article/10.1007/s10462-022-10293-3
24. Orbit Shape Automatic Recognition Based on Artificial Neural Network-Dongmei Du Qing He, ASME 2005 Power Conference http://dx.doi.org/10.1115/PWR2005-50208
25. Fine-grained fault recognition method for shaft orbit of rotary machine based on convolutional neural network- Bo Wu, Songlin Feng, Guodong Sun, Liang Xu, Chenghan
26. Orbit Analysis- Ray Kelm & Dustin Pavelek, Vibration Institute presentation
27. Commercialization of Prognostics Systems Leveraging Commercial Off- The-Shelf Instrumentation, Analysis, and Data Base Technologies -Preston Todd Johnson CBT Technologies

MFPT ANNUAL CONFERENCE

Where Theory meets Practice

CONDITION MONITORING OF RECIPROCATING COMPRESSOR VALVES VIA A STATISTICAL TIME-FREQUENCY APPROACH

Jacob Chesnes[1], Jason Kolodziej[2], Michael W. Anderson[3], and Daniel Nelson[4]
[1,2,3]Rochester Institute of Technology, Rochester, NY, 14623, USA
jjc4939@rit.edu
jrkeme@rit.edu
mwa2912@rit.edu
[4]Novity, Palo Alto, CA, 94304, USA
daniel.nelson@novity.us

ABSTRACT:
The goal of this work is to present a time-frequency-based approach to vibration condition monitoring of gas compressor valves. Due to the cyclostationary nature of reciprocating compressors frequency analysis alone is typically insufficient to assess valve condition. Experimental data is collected on an instrumented Dresser-Rand ESH-1 dual-acting reciprocating compressor by seeding known leakage fault severity levels in the suction and discharge manifold valve assemblies. Fault severity is measured by the leakage hole area and the number of damaged valve elements. By applying a short-time Fourier transform signal processing method to externally mounted vibration measurements, it is possible to identify key regions of interest in the compression cycle. Statistical features are then extracted from these regions and used to train a Bayesian classifier. The presented machine learning approach achieves greater than 90% success in assessing the valve leakage health of the compressor.

KEY WORDS: Condition monitoring; discriminant classification; gas compression technology; time-frequency analysis

INTRODUCTION:

One of the most widely deployed gas compressors used in the petrochemical industry is the reciprocating compressor. It is a very mature technology with over a century of design, manufacturing, and in-service experience. They operate reliably at a wide range of pressures, can compress a large variety of gas, and are highly adaptable thanks to multi-stage capabilities. Unfortunately, reciprocating compressors suffer from higher maintenance costs than centrifugal compressors [1] with the majority of reciprocating compressor downtime and maintenance costs attributed to the valves (36% of shutdowns and 50% of total repair costs). Therefore, condition monitoring of valve assemblies provides a significant opportunity to reduce the overall maintenance cost and provide an economic motivation for condition-based maintenance systems.

The traditional approach to condition monitoring in industry for reciprocating compressors is the analysis of the pressure-volume (P-V) diagram. If the measured P-V diagram deviates from the expected theory certain failure modes like chatter or leakage are predicted. An example of this approach was recently published in 2022 by Chesnes et. al. [2] which presented a polytropic exponent estimation classification method for both the compression and expansion processes. While these diagrams have proven successful they do require the use of real-time, in-cylinder, pressure measurements that add expense and additional maintenance.

Vibration analysis is another common approach that looks for deviations in a machine's typical vibration signature due to a fault condition. Reciprocating machines produce a highly cyclic vibration signature that is represented by a series of periodic events (valve opening and closing, etc.) [3]. Cyclostationary is the term used for this type of vibration and is described as signals that exhibit some periodicity in their energy profile throughout the cycle and have significant variation due to changes in operating and component conditions [4]. Time-frequency transforms in the cycle domain are thus used to capture this cyclostationary nature.

Vibration-based condition monitoring has been the topic of research for valve fault detection in reciprocating compressors for many years and is well presented in a recent 2021 survey paper [5]. Qin et al. [6] showed a method that utilized a denoising step using basis pursuit followed by a novel feature extraction via wave matching and a Support Vector Machine (SVM) classifier for compressor valves with 100% accuracy. Pichler et al. [7] applied a vibration-based spectrogram difference method for compressor operation under varying loads. Valve fault detection was accomplished using SVMs and logistic regression classifiers with high detection accuracy. Potocnik et al. [8] developed a semi-supervised vibration-based condition monitoring method that used principal component and statistical analysis for feature extraction. They then applied several traditional classification methods, such as discriminant analysis, SVMs, and neural networks on real compressor measurements with good performance. Zouari et al. [9] have shown the use of cyclo-stationary modeling for the purposes of reciprocating machine condition monitoring. Regarding valve faults, they identified simple fault indicators through the use of the Wigner-Ville spectrum. Finally, Lin et al. [10], [11] also applied time-frequency analysis of vibration signals with a neural network for automated condition classification and later applied this to valve fault monitoring using seeded faults.

In 2018, Kolodziej & Trout [12], presented a time-frequency domain approach using Short-Term Fourier Transforms (STFT) and the Wigner-Ville Spectrum and through using image processing methods successfully classified seeded valve seat wear and spring fatigue conditions. Next, Scott and Kolodziej [13], extended this work to investigate fault isolation of valve health across all manifolds (suction/discharge), using a novelty detection SVM technique. Finally, Chesnes et. al. [14] further improved the classification success (>90%) by applying a deep-learning approach to compressor valve condition monitoring.

For this original work, a new common fault mode is investigated at the RIT Compression Test Cell, poppet valve leakage, in both the suction and discharge locations at various degrees of severity. Using common operational data (valve cap accelerometers, cylinder pressure, and crankshaft position) a novel feature extraction method is performed to distinguish healthy valves from four levels of degraded (nonhealthy) leaking valves seeded into the ESH-1 compressor. The resulting time-frequency diagrams are processed as images and then used to train a statistical Bayesian classifier to assign a state-of-health classification to the compressor valves.

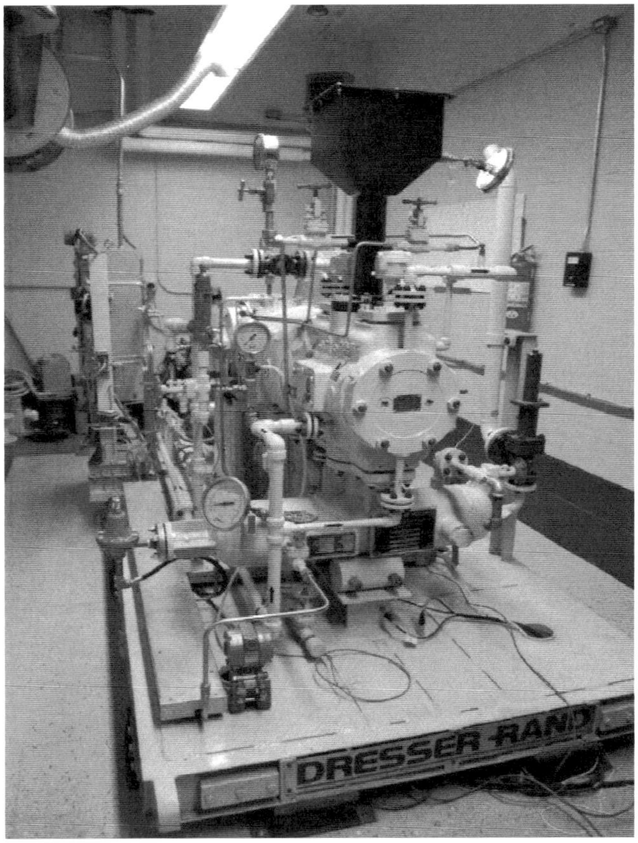

Model:	ESH-1	Max. Temp (°F):	320
Stages:	single	Flow Capacity (ACFM):	34
Piston Diameter:	6"	Compressor (BHP):	7
Stroke:	5"	Weight (lbs):	~8,000
Max. Pressure (psia):	50	Speed (RPM):	360

Figure 1: Dresser-Rand's ESH-1 Reciprocating Compressor located at the RIT Gas Compression Test Cell

COMPRESSOR TEST CELL AT RIT:

Shown in **Figure 1** is the experimental test platform used in this work. It is a Dresser-Rand – A Siemens Business ESH-1 reciprocating compressor located at the Rochester Institute of Technology's (RIT) Compression Test Cell. The ESH-1 is driven by a 10-hp electric motor and has a 6-inch piston with a 5-inch stroke that operates nominally at 360 RPM. The single-stage, dual-acting, compressor, commonly used in the petrochemical industry, was donated by Dresser-Rand and installed at RIT in 2010.

Reciprocating compressors are positive displacement compressors with a piston that pressurizes gas on both sides of the piston, resulting in intermittent airflow. For this work, we denote the piston side closest to the crankshaft as the crank-side cylinder and the piston side furthest from the crankshaft as the head-side cylinder. The compressor operates under full load (both cylinders), half-load (only crank-side), or no load. The image shown in **Figure 2** shows a labeled

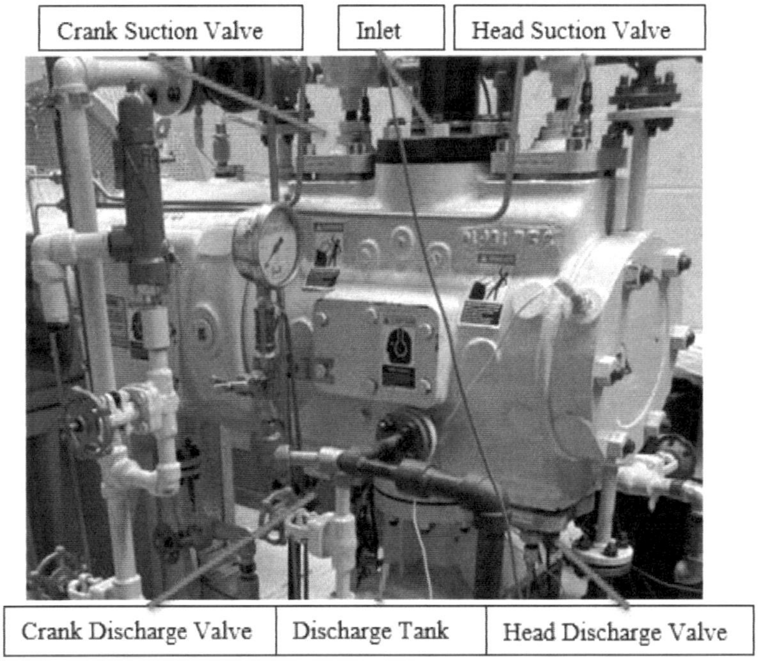

Figure 2: Labeled side view of the Dresser-Rand ESH-1 Reciprocating Compressor

view of each cylinder's set of inlet suction valves that allow air to be drawn in at atmospheric pressure and a set of discharge valves that allow compressed air to be discharged into an anti-pulsation tank. Each valve assembly (**Figure 3**) includes 16 individual poppet valves that are spring-loaded to keep the valves closed until a specific pressure differential is achieved.

Figure 3: The poppet array in the valve assembly

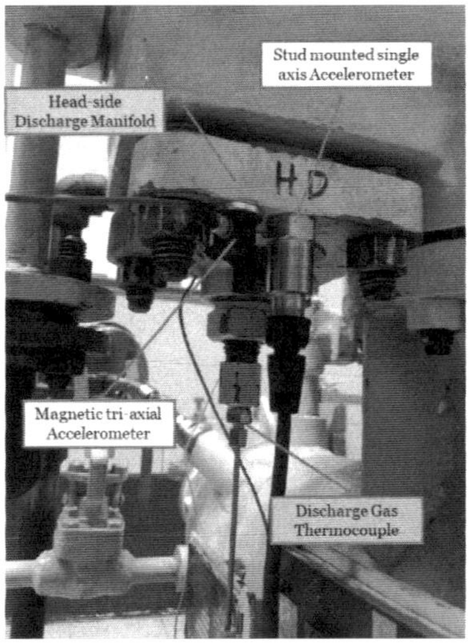

Figure 4: Head-side discharge valve manifold cover with surface mounted accelerometer (PCB 622B01)

Condition monitoring of these valve-spring assemblies is the focus of this research.

Figure 5: A precision machined poppet to provide a leakage path. (1/40" cut in the chamfered edge)

The compression test cell is fully outfitted with a comprehensive NI-based CompactDAQ system to record measurements during compressor operation. The specific sensors utilized for this work are an angular encoder (Photocraft HS20.5QZ) mounted on the main crankshaft, two in-cylinder pressure transducers (Omega PX309-100AI) to measure head and crank-side cylinder pressures, and four single-axis stud-mounted industrial accelerometers (PCB 622B01) permanently affixed to the valve manifold covers (**Figure 4**). Measurements are collected at 12.8 kHz using a custom NI LabVIEW interface to view sensor readings and export data for post-processing.

The main failure mode considered for this work is valve leakage which is a condition that could originate from several root causes such as a crack developing on the edge of one of the poppets due to normal impact fatigue, wear in the chamfered edge in the metal valve assembly overtime preventing a good seal, or even catastrophic poppet failure/destruction due to trapped liquid during the compression stroke. To generate a reliable and quantitative progressive wear condition five levels of seeded leakage wear are created by replacing 0-4 poppets in the assembly with a known and measurable leakage path. **Figure 5** shows the 1/40" cut milled into the chamfered edge of one of the poppets. It is important in any seeded fault testing that the conditions are not random and represents believable scenarios seen in the field under actual run-time operation. While this is not an actual chip in the poppet surface it does provide a consistent measure of leakage when adding multiple poppets in this condition.

METHODOLOGY:

Once the seeded fault conditions are fabricated and measurements collected the next step in the compressor health classification methodology is signal processing. For this work, a series of steps were performed to generate the training and test data for the machine learning classification strategy which is shown in **Figure 6**.

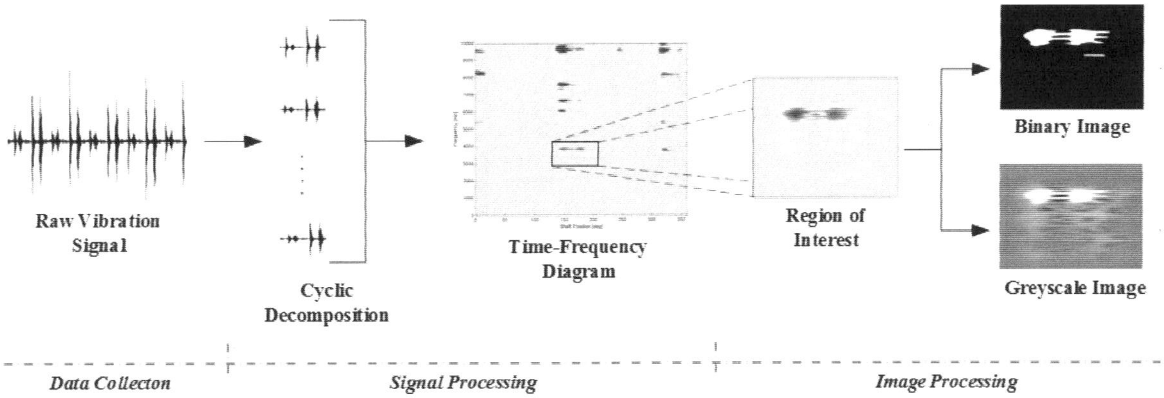

Figure 6: Signal Processing Methodology

First, high-speed data sets are collected for each fault condition consisting of all four valve manifold vibration accelerometers, in-cylinder head and crank-side pressure, and crank angle position as shown in **Figure 7**. Since the crank angle is known, using *cyclic decomposition*, the vibration signal data set is broken into segments, with each segment consisting of an individual compression cycle. Each one of these cycles is treated as a discrete data sample. With a compressor operating speed of 380 RPM, a dataset collected over 50 seconds results in nominally 315 unique samples per valve health condition.

Next, knowing that the compression cycle of a reciprocating compressor is not a stationary process, it is necessary to utilize time-frequency signal processing methods. Several techniques are available for this process, but for this research, the short-term Fourier transform (STFT) is shown in Eqn. (1) and chosen to represent the *time-frequency diagram* (TFD) in **Figure 6** because of its ease of computation and its widespread acceptance. It is noteworthy that the time axis has been transformed to the crank angle position which is important in the next step and is common in rotating machinery. Since it is assumed that the compressor operates at a steady

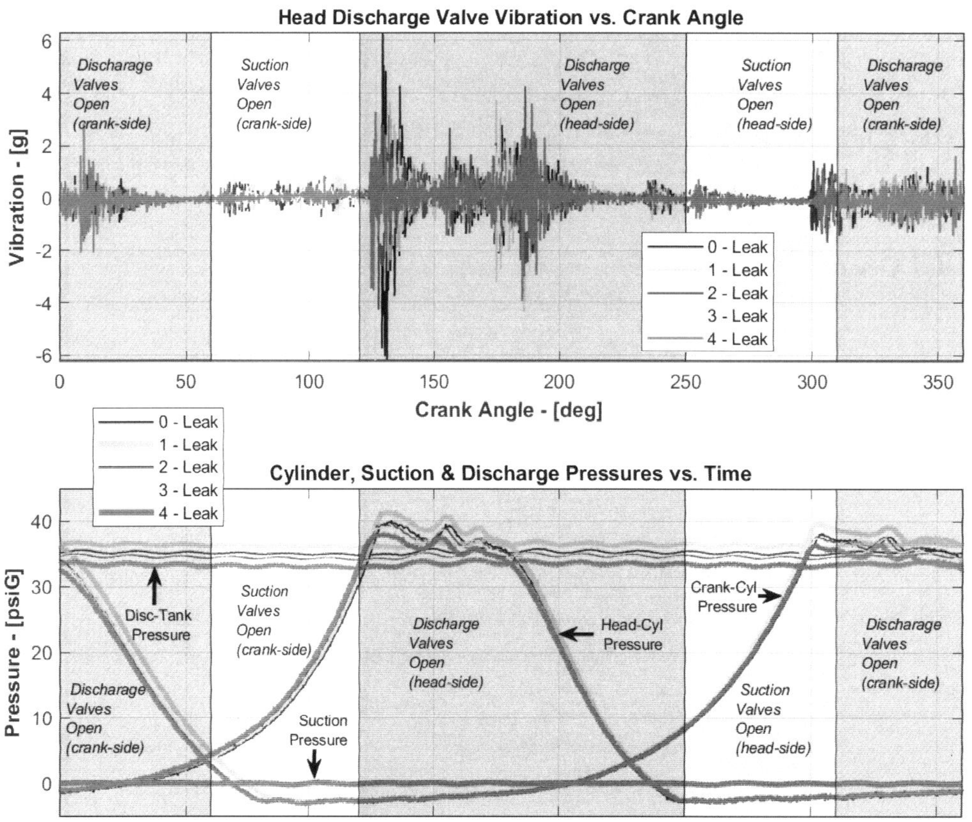

Figure 7: Head Discharge Valve Manifold Vibration vs Crank Angle (top) & Cylinder Pressures vs. Crank Angle

operating speed this conversion is linear and straightforward to implement. There are several options available on how to create this diagram such as balancing the time-frequency resolution trade-off, the amount of overlap, window type, and whether the magnitude scale is logarithmic or not. For this work, a 51-point crank angle window length, 75% overlap, Hanning window, and a linear scale is chosen in an ad-hoc manner due to the success of the overall method but allow for future optimization.

$$STFT(\tau, f_i) = \sum_{k=-\infty}^{\infty} x(k)w(k-\tau)e^{\left(-j2\pi k f_i / N\right)} \tag{1}$$

The next step in the methodology is to extract a targeted region of interest (ROI). This is accomplished by selecting the crank angle range during which the valve is open. In the example shown in **Figure 8,** the head-side discharge valves are under study, and in referring to the cylinder pressure graph in **Figure 7** the crank angle range of valve activity is chosen to be between 125° to 185°. Next, the frequency range is chosen. For this work, the sampling rate fixes the maximum upper limit but the most notable change in magnitude of the TFD during the valve opening event between the different health states is observed between 2,500 and 4,500 Hz which also corresponds to a region less than the bandwidth of the accelerometer. Likewise, the head-side suction valve ROI ranges are chosen and shown in **Table 1**. The range selection process naturally is another opportunity for future optimization of the method.

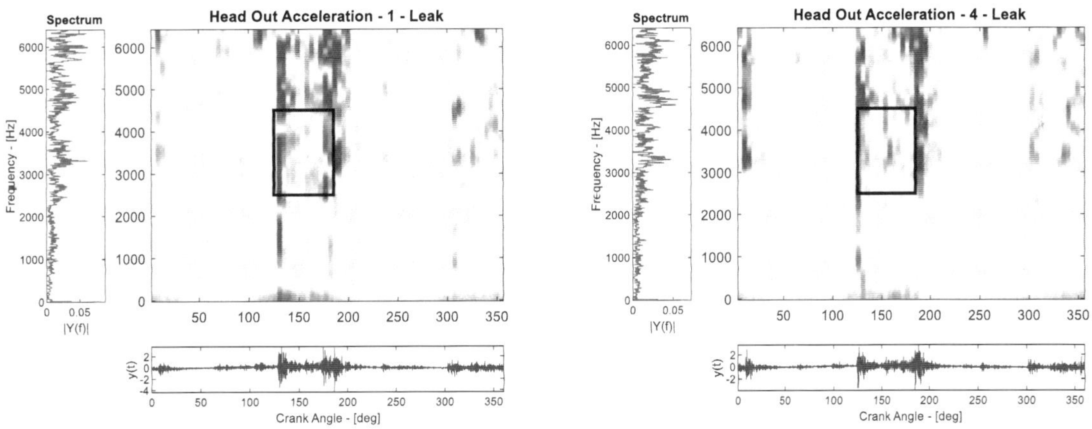

Figure 8: Side-by-Side Comparison of a 1-Leak (left) & 4-Leak (right) in the Head Discharge Assembly Time-Frequency Diagram with the ROI Denoted by the Box.

Table 1: Location of ROIs per Valve Assembly

ROI Location	*Frequency Range (Hz)*		*Crank Angle Range*	
Head Suction	2,500	4,500	240°	290°
Head Discharge	2,500	4,500	125°	185°

Once the ROIs for every cycle have been extracted the next step is to convert them to grayscale and binary images. Similar to the previous steps there were several parameters to be selected for this process. We chose 256 levels of gray for this work with a TFD absolute magnitude range between 0-40. This range was selected to normalize the grayscale range across all health conditions such that the peak (40) is slightly greater than the largest ROI magnitude of any condition collected. To convert from grayscale to binary Otsu's method is used to determine the threshold with holes or objects determined to be too small filled or removed, respectively. **Figure 9** shows one example of this step for the 4-Leak condition. From these two images, a 29-element feature vector is created for <u>each</u> ROI representing 13 first and second-order statistics (e.g. mean, standard deviation, etc.) from the grayscale image and 16 shape features (e.g. area, centroid, etc.) from the binary. A full description of each feature can be found in [12] and calculated using Matlab's Image Processing Toolbox.

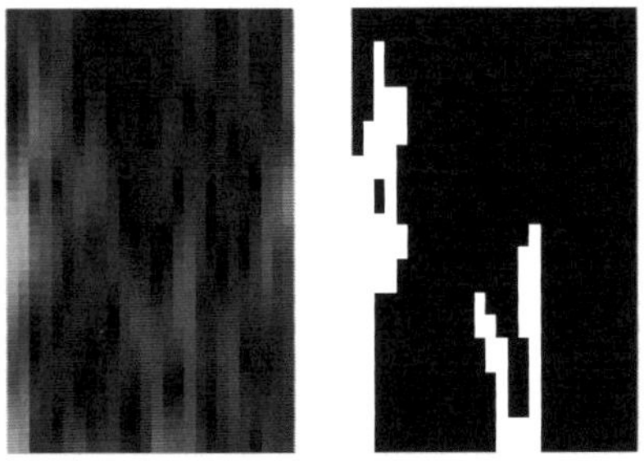

Figure 9: Grayscale and Binary Images of the ROI - 4 Leeks

Finally, to conclude the valve health assessment methodology, each valve leakage level now has a set of unique feature vectors based on individual cycle ROIs. These are used to train linear and quadradic discriminant classifiers in a machine learning sense according to Bayes' theorem to determine decision boundaries. Twenty percent of the samples are randomly reserved to test the accuracy of the trained classifiers. The following section presents all of the scenarios studied in this research using the proposed approach.

RESULTS:

For every test presented in this work, the compressor was running under full load for 3.5 hours to ensure the unit had reached a steady-state operating temperature. All high-speed data (vibration, cylinder pressure, and crank angle) was collected at a 12.8kHz sampling rate every thirty minutes to show temperature-based trends but only the final data set of 50 seconds in duration was used in this analysis. While this unit is commonly used in the petrochemical industry with many types of working fluids, our testing used air as the compressible gas at an inlet temperature of approximately 70°F.

The methodology described above is implemented across several scenarios both on the head-side of the compressor which is under a compression cycle in the 100% loading condition. During the full compression cycle both the discharge and suction valves are actuated and for this work are seeded with a discrete number of leaking poppet valves within the valve manifold as described above (a 1/40" cut against the chamfer of a poppet). Two scenarios were considered for each valve assembly, first zero (healthy), two, and four leaking poppets, and second a more granular five fault severity class arrangement of 0-4 leaking poppets.

Each fault severity condition resulted in 300 unique ROIs (50 seconds of runtime data @ 6.4Hz machine operating speed) for a total of 1500 individual samples (the 3-class scenario had 900 samples) of which 20% was randomly removed from the training process and reserved for testing the trained discriminant classifiers. According to the methodology each ROI is processed as a greyscale and binary image resulting in 13 and 16 features, respectively. To investigate the advantage of adding the binary representation of the ROI, two classifications were performed: a 13-feature, greyscale-only, classifier; then a 29-feature combined greyscale and binary classifier. Due to the ease of implementation after the training feature sets have been defined, both a linear discriminant classifier (LDC) and a quadradic discriminant classifier (QDC) were created.

Table 2: Summary of Valve Classification Results (Training & Test Data) - QDC/LDC

# Classes	# Features	VALVE		QDC	LDC
3	13	Head Suction	Training	81.4%	82.0%
			Test	72.2%	74.4%
5	13	Head Suction	Training	65.8%	63.0%
			Test	60.7%	59.3%
3	13	Head Discharge	Training	88.9%	90.1%
			Test	88.9%	91.1%
5	13	Head Discharge	Training	78.1%	78.0%
			Test	78.0%	78.7%
3	29	Head Suction	Training	91.6%	90.9%
			Test	86.7%	93.3%
5	29	Head Suction	Training	81.1%	78.2%
			Test	70.7%	76.7%
3	29	Head Discharge	Training	93.6%	94.1%
			Test	91.1%	97.8%
5	29	Head Discharge	Training	89.5%	85.2%
			Test	80.7%	84.0%

Table 2 is a summary of all scenarios considered in both valve manifolds. Column 1 separates the 3-class (0, 2, and 4 leaking poppets) from the finer 5-class. The first four rows of the table show the greyscale-only feature vector classifiers followed by the full 29 feature vectors. Column 3 shows the valve test location with Columns 5 and 6 showing the classification results between the LDC and QDCs identifying the training data results from the test data.

While all of the classifications were successful to some degree there were a few that stood out. Overall the *Head Discharge Valve* manifold location met with the strongest classification. We hypothesize that valve motion on the discharge assembly poppets is far stronger than on the suction due to the differential pressure achieved (ΔP = 5psi vs. 3psi) and since vibration sensing is largely based on the impact velocity of the excitation source it is reasonable that discharge manifold accelerometer measurements would show the strongest effect.

Another outcome of the analysis is that adding the greyscale features to the classification strategy primarily improved the weaker suction side valve location. While the discharge manifold had classification accuracies greater than 85% for both the 3-class and 5-class conditions the suction valve leakage condition was particularly poor with only 13 features, especially for the more difficult 5-class scenario (59.5% for the test data). However, adding the additional shape features improved the classification accuracy to 76.7%.

Table 3 - **Table 6** are the corresponding confusion matrices for each condition in **Table 2** where the values in parentheses separate the LDC result from the QDC. These matrices are shown for completeness to further illustrate where the misclassifications occurred. When observing the different performance between LDC and QDC there are instances where the linear classifier outperforms the quadradic. This is probably an anomaly in that the test data set is selected at random and the difference is only a point or two. The decision boundary between the two classifiers is likely very similar. Another aspect of the results that is particularly interesting is that the discharge valve faults are successfully classified at a higher rate than suction valves. From **Figure 7** it is clear that the aggressive fluctuations in the pressure profile during the discharge valve open event quite possibly led to repeated impacts of the poppet on the valve seat. With vibration measurements as the primary input to the classification method, we conclude that this would lead to better class separation and hence a better result. Conversely, the suction pressure profile is relatively stable during the opening event and has few impact occurrences between wear conditions.

Table 3: **Head Suction Valve** – *Test Data Confusion Matrix* – ***13 features***

[top]: 3-class 72.2% QDC & *(74.4% LDC)*

[bottom]: 5-class – 60.7% QDC & *(59.3% LDC)*

Actual \ Predicted	Healthy		2 - Leak		4 – Leak		**Total**
Healthy	26	(27)	1	(2)	2		29
2 – Leak	6	(5)	18	(22)	8	(5)	32
4 – Leak	1	(1)	7	(10)	21	(18)	29
Total	33	(33)	26	(34)	31	(23)	90

Actual \ Predicted	Healthy		1 - Leak		2 - Leak		3 - Leak		4 – Leak		**Total**
Healthy	19	(20)	2	(3)	1	(1)			2		24
1 – Leak	4	(6)	22	(18)	2	(1)	1	(4)	5	(5)	34
2 – Leak	3	(4)	5	(2)	24	(21)	6	(11)	1	(1)	39
3 – Leak	1	(1)	6	(4)	7	(5)	10	(14)	2	(2)	26
4 – Leak			4	(4)	2	(2)	5	(5)	16	(16)	27
Total	27	(31)	39	(31)	36	(30)	22	(34)	26	(24)	150

Table 4: **Head Discharge Valve** – *Test Data Confusion Matrix* – ***13 features***

[top]: <u>3-class</u> 88.9% QDC & *(91.1% LDC)*

[bottom]: <u>5-class</u> – 78.0% QDC & *(78.8% LDC)*

Actual \ Predicted	Healthy	2 - Leak	4 – Leak	**Total**
Healthy	25 *(27)*	5 *(3)*		30
2 – Leak	2 *(3)*	31 *(31)*	3 *(2)*	36
4 – Leak			24 *(24)*	24
Total	27 *(30)*	36 *(34)*	27 *(26)*	90

Actual \ Predicted	Healthy	1 - Leak	2 - Leak	3 - Leak	4 – Leak	**Total**
Healthy	24 *(27)*	1 *(1)*	4 *(1)*		1 *(1)*	30
1 – Leak	*(1)*	24 *(20)*	3 *(6)*	1 *(1)*		28
2 – Leak	5 *(3)*	6 *(5)*	18 *(21)*	1 *(1)*		30
3 – Leak		1 *(1)*	1	26 *(25)*	2 *(4)*	30
4 – Leak		1 *(2)*		6 *(5)*	25 *(25)*	32
Total	29 *(31)*	33 *(29)*	26 *(28)*	34 *(32)*	28 *(30)*	150

Table 5: **Head Suction Valve** – *Test Data Confusion Matrix* – ***29 features***

[top]: <u>3-class</u> 86.7% QDC & *(93.3% LDC)*

[bottom]: <u>5-class</u> – 70.7% QDC & *(76.7% LDC)*

Actual \ Predicted	Healthy	2 - Leak	4 – Leak	**Total**
Healthy	31 (31)			29
2 – Leak		24 (29)	9 (4)	32
4 – Leak		3 (2)	23 (24)	29
Total	31 (31)	27 (31)	32 (28)	90

Actual \ Predicted	Healthy	1 - Leak	2 - Leak	3 - Leak	4 – Leak	**Total**
Healthy	32 (32)					32
1 – Leak		23 (25)		1 (2)	8 (5)	32
2 – Leak		2	22 (22)	5 (7)	5 (5)	34
3 – Leak		(2)	4 (1)	16 (20)	6 (3)	26
4 – Leak		6 (5)	2 (3)	5 (2)	13 (16)	26
Total	32 (32)	31 (32)	28 (26)	27 (31)	32 (29)	150

Table 6: **Head Discharge Valve** – *Test Data Confusion Matrix* – ***29 features***

[top]: <u>3-class</u> – 91.1% QDC & *(97.8% LDC)*

[bottom]*:* <u>5-class</u> – 80.7% QDC & *(84.0% LDC)*

		Predicted			
		Healthy	2 - Leak	4 – Leak	**Total**
Actual	Healthy	23 *(27)*	5 *(1)*		28
	2 – Leak	1 *(1)*	35 *(36)*	1	37
	4 – Leak		1	24 *(25)*	25
	Total	24 *(28)*	41 *(37)*	25 *(25)*	**90**

		Predicted					
		Healthy	1 - Leak	2 - Leak	3 - Leak	4 – Leak	**Total**
Actual	Healthy	23 *(25)*	4 *(3)*	6 *(5)*			33
	1 – Leak		18 *(21)*	2	2 *(1)*		22
	2 – Leak	1 *(1)*	2	23 *(25)*			26
	3 – Leak		*(2)*	1	28 *(29)*	5 *(3)*	34
	4 – Leak	1		1 *(2)*	4 *(7)*	29 *(26)*	35
	Total	25 *(26)*	24 *(26)*	33 *(32)*	34 *(37)*	34 *(29)*	**150**

CONCLUSION:

This research investigated a method to detect reciprocating compressor valve leakage through a non-invasive vibration-based approach. The analysis was performed on the suction and discharge valve assembly of a Dresser-Rand ESH-1 dual-acting compressor using seeded valve faults. Through a series of signal processing steps, the non-stationary nature of the compression cycle was deconstructed to generate statistical image features to train a probabilistic classifier. The resulting validation data show greater than 80% success for five leakage levels and greater than 90% for three.

As with any preliminary research, there are many options for future direction. In this work, faults were only seeded on the valve manifold at a time when in more realistic conditions it is likely faults could occur concurrently. Moreover, valve leakage is only one type of valve condition. Valve seat wear and spring fatigue also are likely to occur as well. Finally, pointed out throughout this paper were several areas that research chose certain options over others such as the frequency range of the ROI and the windowing and overlap on the STFT calculation to name a few. We plan in future research to continue to explore the design space parameters to further improve the accuracy of the compressor valve health assessment.

REFERENCES:

[1] A. G. F. Schirmer, N. F. Fernandes, and J. E. De Caux, "On-line Monitoring of

Reciprocating Compressors," in *NPRA Maintenance Conference*, 2004, pp. 1–20.

[2] J. Chesnes, J. Kolodziej, and D. Nelson, "Statistical Classification Approach to Valve Condition Monitoring Using Pressure Features," *Annu. Conf. PHM Soc.*, vol. 14, no. 1, pp. 1–7, 2022, doi: 10.36001/phmconf.2022.v14i1.3382.

[3] R. B. Randall, *Vibration-based condition monitoring: Industrial, aerospace and automotive applications*. John Wiley, 2011.

[4] J. Antoni, "Cyclostationarity by examples," *Mech. Syst. Signal Process.*, vol. 23, pp. 987–1036, 2009, doi: 10.1016/j.ymssp.2008.10.010.

[5] Q. Lv, X. Yu, H. Ma, J. Ye, W. Wu, and X. Wang, "Applications of machine learning to reciprocating compressor fault diagnosis: A review," *Processes*, vol. 9, no. 6, 2021, doi: 10.3390/pr9060909.

[6] Q. Qin, Z. Jiang, K. Feng, and W. He, "A novel scheme for fault detection of reciprocating compressor valves based on basis pursuit , wave matching and support vector machine," *Measurement*, vol. 45, no. 5, pp. 897–908, 2012, doi: 10.1016/j.measurement.2012.02.005.

[7] K. Pichler, E. Lughofer, M. Pichler, T. Buchegger, E. P. Klement, and M. Huschenbett, "Fault Detection in Reciprocating Compressor Valves for Steady-State Load Conditions," *Mech. Syst. Signal Process.*, vol. 70–71, pp. 104–119, 2016, doi: 10.1016/j.ymssp.2015.09.005.

[8] P. Potočnik and E. Govekar, "Semi-supervised vibration-based classification and condition monitoring of compressors," *Mech. Syst. Signal Process.*, vol. 93, pp. 51–65, 2017, doi: 10.1016/j.ymssp.2017.01.048.

[9] M. W. R. Zouari, J. Antoni, J.L Ille, M. Sidahmed, M. Milaert, "Cyclostationary Modelling of Reciprocating Compressors and Application to Valve Fault Detection," *Int. J. Acoust. Vib.*, vol. 12, no. 3, pp. 116–124, 2007.

[10] Y.-H. Lin, H.-C. Wu, and C.-Y. Wu, "Automated condition classification of a reciprocating compressor using time–frequency analysis and an artificial neural network," *Smart Mater. Struct.*, vol. 15, no. 6, pp. 1576–1584, Dec. 2006, doi: 10.1088/0964-1726/15/6/009.

[11] Y.-H. Lin, H.-S. Liu, and C.-Y. Wu, "Automated valve condition classification of a reciprocating compressor with seeded faults: experimentation and validation of classification strategy," *Smart Mater. Struct.*, vol. 18, no. 9, p. 095020, Sep. 2009, doi: 10.1088/0964-1726/18/9/095020.

[12] J. R. Kolodziej and J. N. Trout, "An image-based pattern recognition approach to condition monitoring of reciprocating compressor valves," *JVC/Journal Vib. Control*, vol. 24, no. 19, pp. 4433–4448, 2018, doi: 10.1177/1077546317726453.

[13] C. M. Scott and J. R. Kolodziej, "A novelty detection approach to compressor valve fault

detection and localization," *ASME Int. Mech. Eng. Congr. Expo. Proc.*, vol. 14, pp. 1–10, 2020, doi: 10.1115/IMECE2020-23314.

[14] J. Chesnes and J. Kolodziej, "Application Based Comparison of Statistical Versus Deep Learning Approaches To Reciprocating Compressor Valve Condition Monitoring," *Annu. Conf. PHM Soc.*, vol. 13, no. 1, pp. 1–9, 2021, doi: 10.36001/phmconf.2021.v13i1.3081.

Jacob J. Chesnes is currently pursuing a Ph.D. in Mechanical and Industrial Engineering at the Rochester Institute of Technology (RIT) in Rochester, NY with expected completion in 2024. He received a B.S. in Mechanical Engineering and an M.E. in Mechanical Engineering from RIT in 2021. He has been a production support engineer co-op at G. W. Lisk tasked with decreasing the rate of solenoid valve failure. His most recent project was to design an inclusive retractable game net easily set up on standard cones for Inclusivity Inc.

Jason R. Kolodziej is an Associate Professor of Mechanical Engineering at the Rochester Institute of Technology (RIT) in Rochester, NY. He received his Ph.D. in mechanical engineering from the State University of New York at Buffalo in 2001 with a research focus on controls and nonlinear system identification. For eight years he worked in industry for General Motors Fuel Cell Activities as a Sr. Research Engineer with principal duties in hybrid electric-fuel cell vehicle powertrain controls and system architecture. To date, he has been granted 10 U.S. Patents. His present research focus is the study of fault detection, diagnosis, and prognostic health assessment of engineering systems. He currently has funded projects covering a wide range of industrial applications from electromechanical actuators in aircraft to fuel cell automotive powertrains to large-scale compression equipment. He is a member of the ASME. In 2012, he was awarded RITs prestigious Eisenhart Provost Award for Excellence in Teaching.

MICHAEL GOES HERE

Daniel Nelson has over 10 years of experience in high-fidelity modeling and numerical methods, with particular emphasis on fluid dynamics and tribo-mechanical systems. At Novity, he applies his expertise to the development of physics-based algorithms, including model-based reasoning, physics of failure, and prognostics and diagnostics of industrial machinery. Before Novity, Dr. Nelson has worked for several OEMs in the oil and gas, chemical processing, and aerospace industries, where his focus was on multi-physics modeling of tribological elements and gas compression equipment. Dr. Nelson earned his Ph.D. at U.C. San Diego, where he studied computational fluid dynamics and dynamical systems, and his B.S. in Aerospace Engineering at San Diego State University.

MFPT ANNUAL CONFERENCE
Where Theory meets Practice

SEPARATION OF VIBRATORY COMPONENTS IN COMPLEX SYSTEMS FOR CONDITION MONITORING

F. Karkafi [(1), (2), *], D. Abboud [(1)], Q. Leclère [(2)], J. Antoni [(2)], M. El Badaoui [(1), (3)]

[(1)] Safran Tech, Rue des Jeunes Bois – Châteaufort 78772 Magny–les–Hameaux, France
[(2)] Univ Lyon, INSA Lyon, LVA, EA677, 69621 Villeurbanne, France
[(3)] Univ Lyon, UJM-St-Etienne, LASPI, EA3059, F-42023, Saint-Etienne, France

[*]*Corresponding author:*
E-mail address: fadi.karkafi@insa-lyon.fr

ABSTRACT:
Mechanical signals are often a mixture of multiple components produced by different sources within multiple subsystems. The separation of these components is beneficial for differential diagnosis, fault severity assessment and prognosis. The synchronous average (SA) is a powerful and widely used technique for this purpose, providing optimal estimation performance and simplicity. However, the complexity of such systems often implies the existence of several spectral interferences, which jeopardize the straightforward application of the SA. This paper addresses this issue by proposing a general strategy based on the synchronous average and the set theory, to isolate each vibratory contribution related to a source of interest, while accounting for spectral interferences. This approach enables a comprehensive understanding of the system's health status and the ability to locate specific issues within a subsystem. The proposed methodology is demonstrated through an example of an accessory gearbox vibration signal acquired during a ground test campaign on a CFM56 turbojet engine.

KEY WORDS: Synchronous Average; Mechanical Contribution Separation; Condition Monitoring; Vibration Analysis.

I. INTRODUCTION:

The study of mechanical systems has long been a crucial area of research in various industries [1]. Effective monitoring and diagnosis of mechanical systems is essential for maintaining smooth operations. Vibration and acoustic analysis is a widely used method for monitoring the health of mechanical systems. This approach is non-intrusive and provides valuable information about the internal forces in the system, offering a reliable tool for fault detection and diagnosis [2]. However, these systems are often complex, involving the conversion of multiple speed and torque ratios to achieve different tasks. Therefore, this often results in a large number of rotating parts, each of which contributes differently to the overall vibration signal. This makes it difficult to identify and diagnose specific components, thus the importance of signal separation in [3]. Furthermore, the presence of multiple subsystems and rotating parts can lead to some interferences between different sources. This problem is often addressed through frequency analysis, which assumes that the sources do not share their frequency content. However, in many real-world applications, frequency content overlap is unavoidable, which results in a spectral interference due to having the information of a certain line contribution masking the other. This superimposition can compromise the accuracy of diagnosis in terms of fault detection and localization by tracking the faulty component over the healthy one and vice versa, therefore, a false health state assumption. Consequently, to diagnose each mechanical component independently, it is essential to extract the exclusive contribution of each one by eliminating the overlapped frequencies caused by other components.

Some generic signal processing techniques were proposed for signal decomposition and mechanical component contribution extraction using vibration analysis. Starting with a basic Synchronous Average (SA); this method involves averaging the signal over one or several periods to extract the sinusoidal component that refers to a certain frequency and its multiples. The advantage of this method is its simplicity making it possible to implement without any estimation. However, when considering a set of lines, the sum of all contributions would result in a replication of the interference mentioned before. This concept was addressed in [4] by exploiting the corresponding operator of the synchronous average and this replication problem when extracting the whole subsystem. However, the remaining spectral overlap for each contribution still exists which triggers again the problem stated before. Next, [5] involves the use of a band pass filter to extract the frequency content of the components of interest. This method is simple to implement and can provide good results for simple mechanical systems. However, it can be sensitive to the choice of filter parameters and may include some interferences within the chosen bandwidth around the centered frequency. In addition, the problem of frequency overlap is still considered having several sources sharing the same frequency. Otherwise, the Optimal Subspace-Kurtosis and Subspace-Extended Kalman Filter involves combining the advantages of both approaches to achieve optimal source separation [6]. It uses the subspace-kurtosis method to extract periodic signals of interest, while the extended Kalman filter is used to estimate the corresponding contributions. The advantage of this method is that it provides a comprehensive solution to the problem of interferences and source separation, taking into account both the presence of periodic signals and the correction of interferences. Other works use Kalman filters to improve frequency analysis and order tracking methods [7, 8, 9 ,10 ,11] and recent techniques aim to improve the Vold Kalman Filtering for resampling techniques, high frequency resolution in Fourier analysis and crossing orders detection [12, 13, 14, 15, 16]. However, when many harmonics are considered in typical complex mechanical systems, it can be challenging for these approaches to accurately track all of them, which can lead to an increased computation time and cost.

In this paper, the proposed method addresses the challenge of interferences and contribution separation in complex mechanical systems. In Section 2, the problem statement defines the periodicity identification in such systems and its replication when extracting several sources. Section 3 presents the methodology of the proposed method systematically to solve the defined problem. Section 4 demonstrates the application of the method on an aircraft accessory gearbox. Finally, in Section 5, a conclusion wraps this paper up and highlights the most important details.

II. PROBLEM STATEMENT:

Vibratory signals generated by rotating machines are periodic in nature. This periodicity is directly linked to the angle of rotation of the machine. In fact, when machines operate under a perfectly constant speed, this periodicity remains consistent over time. In practice, machines are subject to speed fluctuations, which can introduce errors when analyzing time domain signals. One widely used solution in mechanical signal processing is angular description [17, 18]. This can be achieved by resampling the time domain signal in the angular domain using an encoder signal indicating the position of the angle of rotation.

In this context, the synchronous average (SA) is a powerful and widely used tool in vibration analysis, which particularly deals with the extraction of periodic waveforms (such as produced by imbalances, misalignments, anisotropic rotors, flexible coupling, gear meshing and other phenomena) from other competing but random sources, assuming that its period is known a priori [19, 20, 21]. The basic idea of the SA is to average the signal over a number of samples called a cycle. This is done by dividing the signal into segments, each corresponding to one cycle of the signal, and then averaging the values within each segment. Mathematically, let $x(\theta)$ be an angular signal of cycle Θ (for instance $\Theta = 2\pi$); its SA is defined as follows:

$$SA_\alpha(\bar{\theta}) = \langle x\left(\bar{\theta} + m\frac{2\pi}{\alpha}\right)\rangle_m, \quad m \in \mathbb{N} \tag{1}$$

where the operator $\langle f(m)\rangle_m$ denotes the empirical average of the function $f(m)$, α being the fundamental cyclic order (so that the averaging period equals $2\pi/\alpha$) and $\bar{\theta} \in [0, \frac{2\pi}{\alpha}]$ for the angular location in the cycle (i.e. $\bar{\theta} = \theta - k\frac{2\pi}{\alpha}$, with k the greatest integer smaller than $\theta / \frac{2\pi}{\alpha}$).

The obtained SA signal is defined over a corresponding period with respect to a cyclic order α and then replicated to match with initial signal length as follows:

$$m_\alpha(\theta) \leftrightarrow SA_\alpha(\bar{\theta}) \tag{2}$$

The SA is typically used to extract the vibratory component associated with a rotating shaft, resulting in a Dirac comb in the frequency domain at the line order and its multiple harmonics. The order of each shaft is defined as the ratio between the frequency of the shaft and the reference frequency, with the reference shaft having an order of 1.

When approaching the problem of a complex mechanical system decomposition, given a general model illustrated in Figure 1 with the corresponding interferences in red, it is essential to firstly isolate the subsystem of interest (a_j, such that $j = 1, \dots, N$). Some examples of subsystems in mechanical systems are: an accessory gearbox (AGB), a gas generator, a combustion chamber, etc…

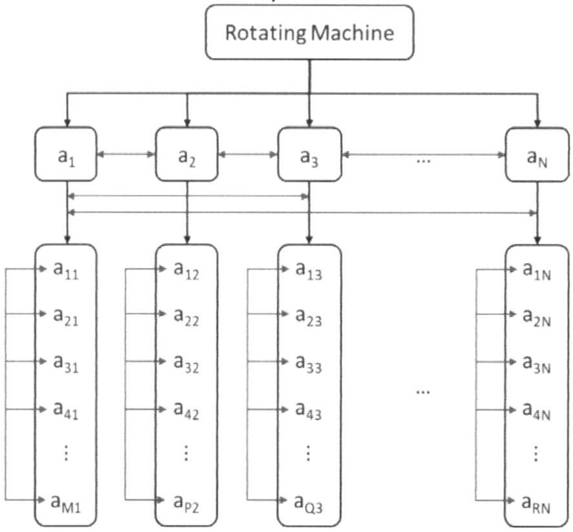

Figure 1. General model of a rotating machine considering different interferences (red arrows) between different lines (a_{11}, a_{12}, ...) in each subsystem (a_1, a_2, ...)

Once the isolation of the potentially defective subsystem has been carried out, it is important in a second phase to proceed with the separation of the various components linked to the rotating parts of this subsystem (i.e. a_{i1}, such that $i = 1, \dots, M$, from the subsystem a_1). This separation considers the exclusive contribution of each component (without interferences) to accurately locate the fault in the isolated subsystem.

In the following, and without loss of generality, the problem addressed in this paper is explained through a simplistic example. Considering a gearbox with an input shaft rotating at order 1 (reference shaft) and an output shaft rotating at 1.2, the associated frequency set with the SA of each shaft would lead to the following order set:

Input Shaft: S_i: $\{\alpha_i = 1; 2; 3; 4; 5; 6; \dots$, such that $\alpha_i \leq \frac{N_S}{2} \in \mathbb{N}\}$
Output Shaft: S_o: $\{1.2 * k = 1.2; 2.4; 3.6; 4.8; 6; \dots$, such that $1.2 * k \leq \frac{N_S}{2} \;/\; k \in \mathbb{N}\}$

with N_S being the number of samples per reference cycle assumed 2π in this paper.

The vibratory contribution of this gearbox (the subsystem) cannot be directly obtained by summing up the SA of both shafts. In fact, as seen in the Figures 2a – 2b, there are several orders in common between the shafts (i.e. 6 and its multiples 12, 18, 24, etc.). These common orders represent the frequency interferences of the subsystem components. The sum of both SA leads to incorrect estimation, which manifests through higher energy at the shared frequencies in the spectrum (see in Figures 2c – 2d). To be noted that this interference is not meant to be completely removed for the subsystem but rather its replication because it can provide important information about its global health state. In fact, these shared harmonics, representing the meshing harmonics and their energies, are widely used as efficient diagnostic indicators in gearboxes.

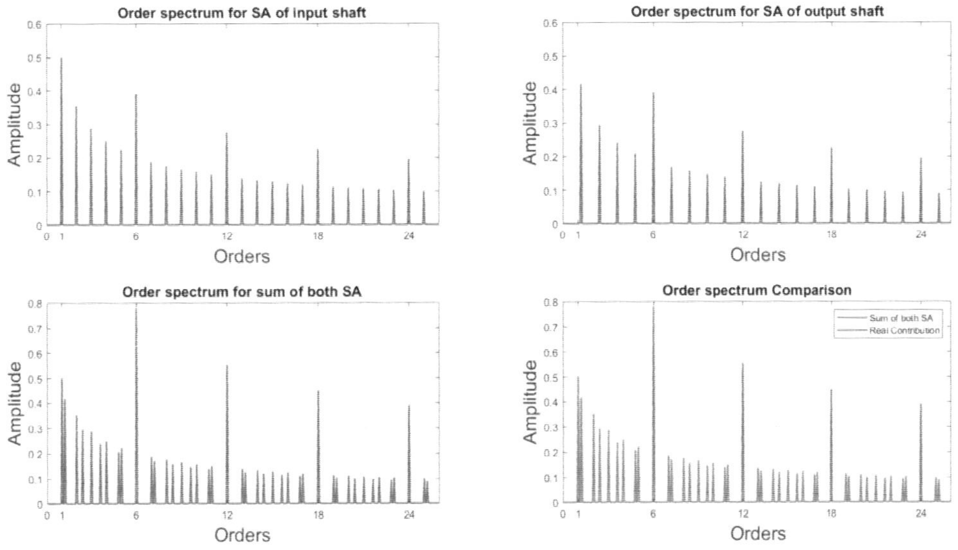

Figure 2. Order spectra for:

a- (Top left) SA of input shaft,

b- (Top right) SA of output shaft,

c- (Bottom left) Sum of both SAs,

d- (Bottom right) Comparison between the sum of both SAs and the real subsystem SA

Another interesting issue addressed in this paper is the localization of a fault. This needs to be achieved by considering the exclusive contribution of each shaft so the faulty one does not affect the others. As seen in Figures 2a – 2b, the SAs of the two shafts present an interference at the sixth order and its multiples, which could mislead the identification of the faulty shaft. Therefore, it is essential to eliminate the remaining interferences between each source to evaluate the health indicator applied on the exclusive contribution of each one. This paper goes beyond this simple example to the case of more complex systems, where multiple shafts with different rotating frequencies having multiple common frequencies in play.

III. PROPOSED METHODOLOGY:

The proposed methodology in this paper is designed to isolate the desired vibratory contribution of a given subsystem in a complex mechanical system using the SA, as well as the exclusive vibration response of each of the subset component. On the one hand, this is achieved by using the set theory to identify the common frequencies that represent the interferences between each line and eliminate its replication as seen in [4]. On the other hand, it can be implemented by the computation of the least common multiple (LCM) of each pair of orders.

Figure 3 illustrates this theory, where multiple sets are considered, having the set "a_1" representing the reference shaft. Each set corresponds to the orders ensemble of each shaft's contribution resulted from its SA. By comparing the orders of shafts a_1 and a_2, it can be seen that some interferences exist (hatched part) between these two sets and others that are jointly common with shaft a_3 (green triangles).

Subsystem

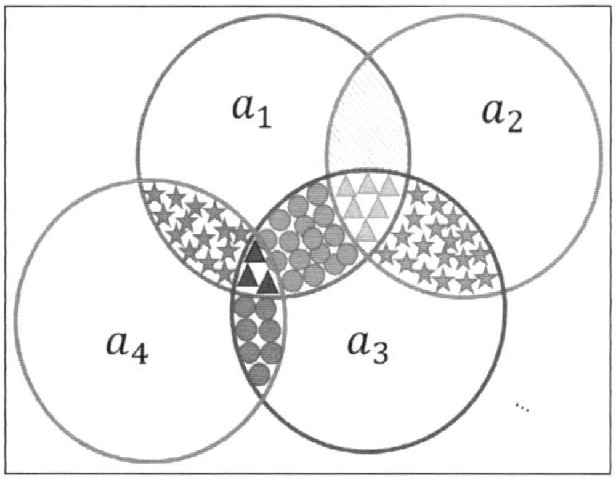

Figure 3. Illustration of the sets of SA orders with the different interferences

Generally, one can consider N shafts with the corresponding set of fundamental orders $\mathcal{A} = \{\alpha_1, \alpha_2, \alpha_3, \dots, \alpha_N\}$ constituting a certain subsystem. The goal is to find the signal associated with this specific subsystem, which encompasses the total set of constituent lines excluding the replications of the interference between each considered line.

This leads to the following equation:

$$\sum_{\alpha_i \in \mathcal{A}} m_{\alpha_i}(\theta) = s(\theta) + \Delta(\theta), \tag{3}$$

with $m_{\alpha_i}(\theta)$ being the SA of the angular signal $x(\theta)$ with respect to the order α_i, $s(\theta)$ being the subsystem deterministic signal and $\Delta(\theta)$ being the replication contribution caused by the interfering components.

Taking the replication to be the unique interference between each two sets (i.e. $a_1 \cap a_2$ without $a_2 \cap a_1$); the interference is not totally eliminated between these two sets, but rather the replication of the interference.

The replication set requires a more generic definition because common orders between physically unlinked shafts can still exist. In addition, this interference will repeat for integer multiples of the common order, leading to two major problems that prompted the proposal of two solutions:

- Calculate the LCM between each pair of shafts to find the fundamental common order.
- Compute the synchronous average relative to the LCM order found to take into consideration all its multiples.

In order to calculate the corresponding LCM of all common orders for each shaft, a matrix of common orders (MCO) is proposed. To be able to free the replication of the interference between each two shafts without the elimination of the latter, this matrix must be an upper triangular matrix whose diagonal corresponds to the common order between two consecutive shafts. By taking the LCM between each pair of shafts, only the lowest common order is considered, eliminating the redundant frequencies when isolating the subsystem.

Thus, the computation of the following matrix of common orders (MCO):

$$\mathcal{M}(i,j) = \begin{cases} LCM(\alpha_i, \alpha_{j+1}), & if\ i \le j \\ 0 \quad , & if\ i > j \end{cases} \tag{4}$$

with $LCM(a,b) = \frac{a*b}{GCD(a,b)}$ and i and j iterating from $1 \to N-1$, therefore $\mathcal{M}$ would be of size $(N-1, N-1)$.

The greatest common divisor (GCD) can be computed using several methods, i.e. the Euclidean algorithm that considers a kth step that performs a division with remainder to find the quotient q_k and remainder r_k so that : $r_{k-2} = q_k r_{k-1} + r_k$ with $r_{k-1} > r_k \ge 0$. In the initial step k = 0, the remainders are set to $r_{-2} = a$ and $r_{-1} = b$, thus the algorithm must eventually produce a zero remainder $r_N = 0$. Therefore, the final nonzero remainder is the greatest common divisor of a and b: $GCD(a,b) = r_{N-1}$.

Once the MCO is calculated, it is used to find the SAs with respect to each found order and subtract it from the total sum of SA of all considered shafts. This ensures that the final result contains a single interference between each shaft, without modifying the structure of the total set. In conclusion, the formula (3) becomes:

$$s(\theta) = \sum_{\alpha_i \in \mathcal{A}} m_{\alpha_i}(\theta) - \sum_i \sum_j m_{\mathcal{M}(i,j)}(\theta), \tag{5}$$

with $m_{\mathcal{M}(i,j)}(\theta)$ being the SA of the angular signal $x(\theta)$ with respect to the order found at element $\mathcal{M}(i,j)$.

Furthermore, to extract the exclusive contributions of each shaft from the isolated subsystem, the same set theory and the matrix MCO are used to find the following:

$$s(\theta) = \sum_{\alpha_i \in \mathcal{A}} \left(c_{\alpha_i}(\theta) + \varepsilon_{\alpha_i}(\theta)\right), \tag{6}$$

with $c_{\alpha_i}(\theta)$ being the exclusive contribution of shaft of order α_i. The remaining interference $\varepsilon_{\alpha_i}(\theta)$ represents all common contributions between the shaft of order α_i and all remaining ones as seen in the following:

$$\varepsilon_{\alpha_i}(\theta) = \sum_j s_{\mathcal{M}(i,j)}(\theta), \tag{7}$$

with $s_{\mathcal{M}(i,j)}(\theta)$ being the SA of the subsystem deterministic signal $s(\theta)$ with respect to the order found at element $\mathcal{M}(i,j)$ and i being the index of α_i in the set $\mathcal{A}$. One can identify the following:

$$c_{\alpha_i}(\theta) = s_{\alpha_i}(\theta) - \varepsilon_{\alpha_i}(\theta), \tag{8}$$

with $s_{\alpha_i}(\theta)$ being the SA of the subsystem deterministic signal $s(\theta)$ with respect to order α_i.

In conclusion, each shaft's exclusive contribution is extracted as follows:

$$c_{\alpha_i}(\theta) = s_{\alpha_i}(\theta) - \sum_j s_{\mathcal{M}(i,j)}(\theta), \tag{9}$$

In summary, the proposed methodology in this paper uses the set theory and the calculation of MCO to isolate each subsystem within a complex mechanical system. By eliminating the redundant frequencies present in the synchronous averages of the shafts' orders, one can finally obtain the exclusive contribution of each component.

IV. APPLICATION:

In this section, an experimental example is considered to understand how to apply such method on aircraft subsystems. It concerns a vibration signal acquired during a ground test campaign on the CFM56 turbojet engine (Figure 4), which is a civil aircraft engine with two main shafts low and high pressure, and an accessory gearbox linked to the high-pressure shaft. The acquired signal is obtained from the accelerometer called Acc2 with a sampling frequency of 50,000 Hertz and a time duration of 200 seconds. The acquisition was performed during a steady state at idle followed by a slow acceleration up to full power. For this application, the desired subsystem to isolate is the accessory gearbox. The kinematics of the gearbox and the rotating speeds of its shafts are described in Figure 5. The corresponding signal provided by the Acc2 accelerometer (located near the shafts L4 and L5 of the AGB) and the rotating frequency of shaft L1 is depicted on Figure 6. This signal has been initially provided for a contest during the International Conference Surveillance 8, October 20 and 21, 2015, at the Roanne Institute of technology, France. More details about the signal can be found in [22].

Consequently, the orders are structured in a well-designed table to clearly distinguish the order of each shaft. A table was provided during the contest for the same specific reason by listing all characteristic frequencies of the gearbox components. In this embodiment, 12 frequencies are given and associated with the rotation of the 12 shaft lines of the CFM56 AGB. Table 1 summarizes the dynamics of the AGB with the 12 frequencies of the set $f_{reference}$ of shaft frequencies referenced by N_2. It is specified that these frequencies are obtained with a rotational speed of shaft N2 of 10,000 rpm.

MFPT ANNUAL CONFERENCE
Where Theory meets Practice

The orders considered in this kinematics correspond to the Tooth line F_1 (N_2). Consequently, the goal is to find all the interferences between the HP shaft and all the others, then RDS and all the others, etc. In addition, it is necessary to analyze the different sources that interfere jointly (an interference might exist for several sources even if they are not physically linked).

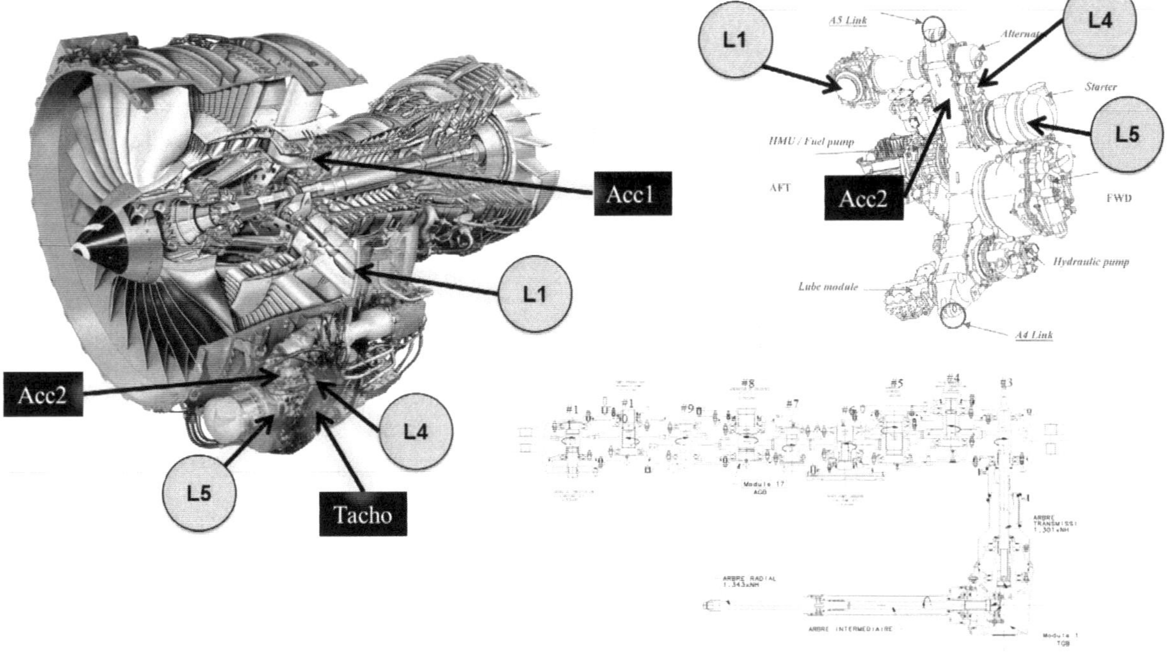

Figure 4. General overview of the engine and the accessory gearbox.

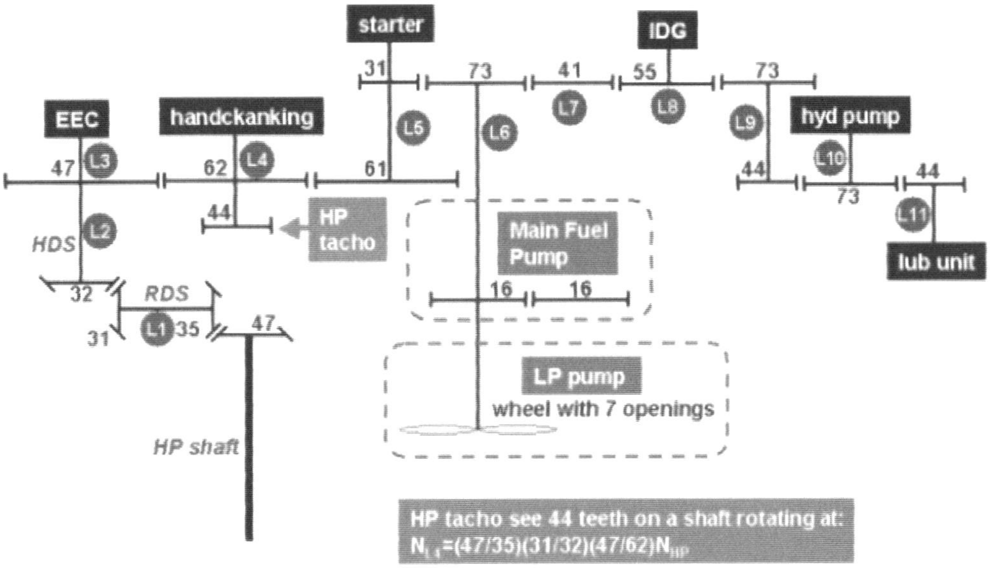

Figure 5. Kinematics of the gearbox.

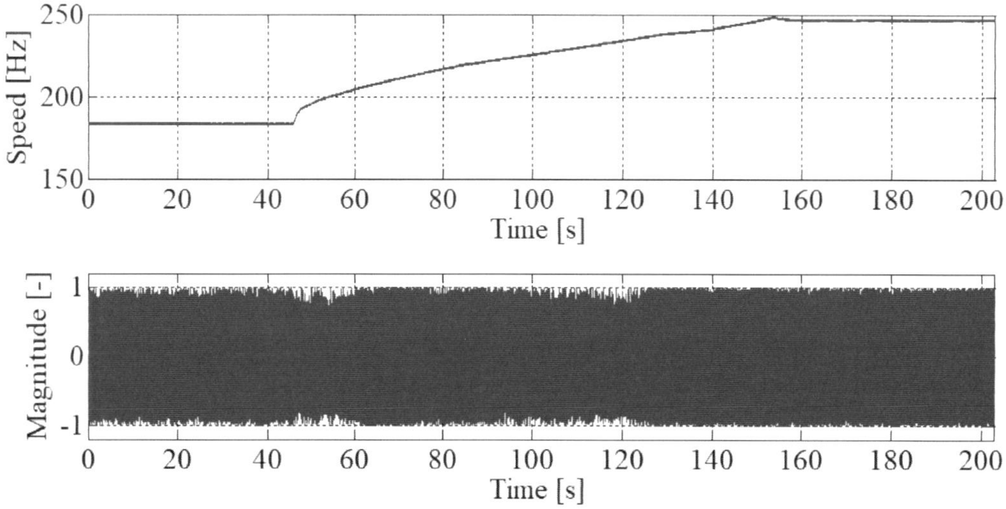

Figure 6. (Top) Evolution of the gas generator shaft speed N1 in Hertz and (bottom) the raw vibration signal measured on the gas generator casing

Table 1. Characteristic frequencies of the gearbox

N2 speed (rpm) 10000	Line 1		Line 2		Line 3		N2 tacho: Line 4		Line 5		Line 6		Line 7		Line 8		Line 9		Line 10		Line 11	
	HP Shaft	RDS		Horizontal Drive Shaft		EEC Alternator		Handcranking		Starter		Fuel Pump		Intermediate		IDG		Intermediate		Hydro Pump		Lubrication Unit
	aval	amont	aval	amont	aval	amont	aval	amont	aval	amont	aval	amont	aval	amont	aval	amont	aval	amont	aval	amont	aval	amont
Number of teeth	47	35	31	32	47	47	47	62	62	61	31	73	73	41	41	55	55	73	44	73	73	44
Rotating Speed (rpm)	10000	13429		13009		13009		9862		10023		4256		7579		5649		4256		2566		4256
F1 tooth (Hz)	167	224	224	217	217	217	217	164	164	167	167	71	71	126	126	94	94	71	71	43	43	71
F1 tooth (N2)	1.0000	1.3429	1.3429	1.3009	1.3009	1.3009	1.3009	0.9862	0.9862	1.0023	1.0023	0.4256	0.4256	0.7579	0.7579	0.5649	0.5649	0.4256	0.4256	0.2566	0.2566	0.4256
Gear Mesh Frequency (Hz)	7833		6938		10190		10190		10190		5179		5179		5179		5179		3121		3121	
Gear Mesh (mult N2)	47.0000		41.6286		61.1420		61.1420		61.1420		31.0721		31.0721		31.0721		31.0721		18.7284		18.7284	
F coincidence (Hz)	4.8		7.0		216.8		3.5		2.7		2.3		1.7		2.3		1.3		1.0		1.0	

Consequently, the next step would be to calculate the MCO with the corresponding orders. Figure 7 shows the resulting matrix after including all orders provided by the Table 1. It is essential to note that a non-integral number is mostly anticipated for each order (aside the reference one). However, the synchronous average takes an integer number of samples as the period to average the signal. As a result, it is possible to find a common order between each one.

α_i \ α_{j+1}	1.3429	1.3009	1.3009	0.9862	1.0023	0.4256	0.7579	0.5649	0.4256	0.2566	0.4256
1.0000	47.0000	147.0000	147.0000	285.0000	1.0000	63.0000	72.0000	274.0000	63.0000	137.0000	63.0000
1.3429	0	41.6286	41.6286	63.1143	209.4857	77.8857	76.5429	256.4857	77.8857	22.8286	77.8857
1.3009	0	0	1.3009	61.1420	61.1420	300.5062	165.2134	98.8679	300.5062	92.3634	300.5062
1.3009	0	0	0	61.1420	61.1420	300.5062	165.2134	98.8679	300.5062	92.3634	300.5062
0.9862	0	0	0	0	61.1420	80.8652	553.2362	54.2388	80.8652	233.7201	80.8652
1.0023	0	0	0	0	0	31.0721	31.0721	31.0721	31.0721	11.0256	31.0721
0.4256	0	0	0	0	0	0	31.0721	31.0721	0.4256	18.7284	0.4256
0.7579	0	0	0	0	0	0	0	31.0721	31.0721	65.9336	31.0721
0.5649	0	0	0	0	0	0	0	0	31.0721	109.0350	31.0721
0.4256	0	0	0	0	0	0	0	0	0	18.7284	0.4256
0.2566	0	0	0	0	0	0	0	0	0	0	18.7284

Figure 7. Resulting MCO values (in black) with respect to each order (in blue)

It can be identified in the matrix above that the diagonal presents the meshes of the gears located on the consecutive shafts or a divisor of this mesh in case a LCM smaller than the mesh was found. Afterwards, using the given instantaneous angular speed, the signal undergoes resampling into its angular domain having the corresponding trajectory with its order spectrum in Figure 8a. The first phase of the proposed method extracts the contribution of the AGB by utilizing MCO to eliminate the replication of interferences among the subsystem lines. As a result, Figures 8b and 8c depict the signals of the AGB (deterministic component) and the residual, along with each corresponding order spectra. This deterministic signal corresponds to the periodic part of all the shafts of the gearbox including the common parts (the interferences) between the different lines without any replication.

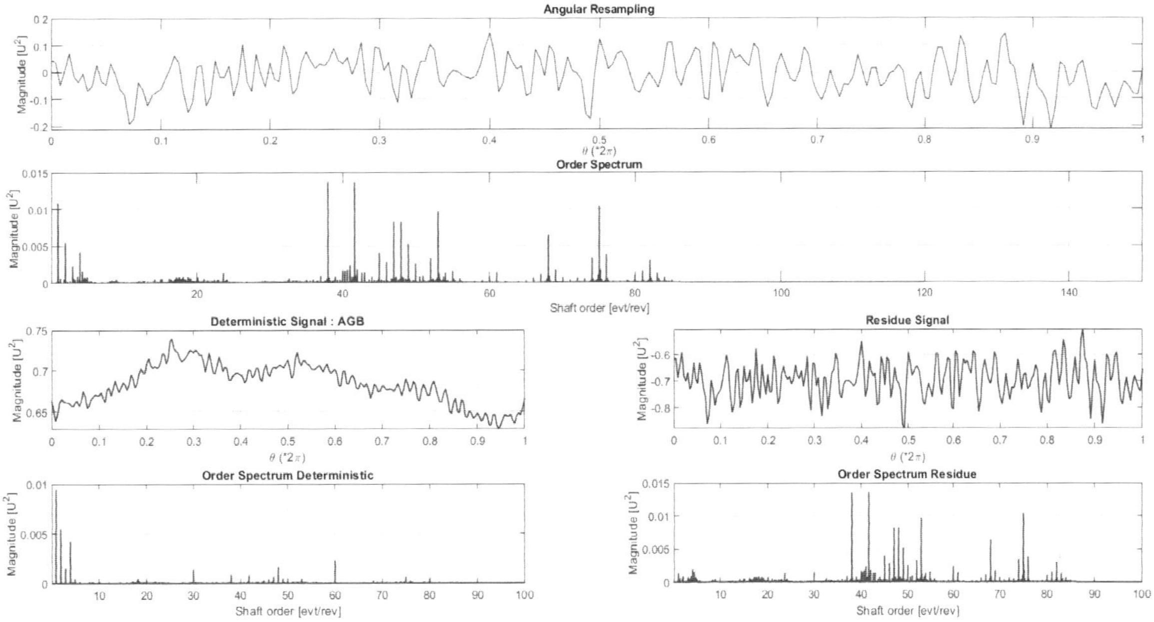

Figure 8.
a – (Top Center) CFM56 angular signal and its order spectra
b - (Bottom Left) Exclusive CFM56 AGB deterministic Signal and its order spectra
c - (Bottom Right) CFM56 signal without AGB and its order spectra

Hence obtaining the contribution of each shaft without any remaining interference using the second phase of the proposed approach with the same MCO computed before. Regarding the 12 lines constituting the AGB, the different SAs are visualized in Figure 9 and their order spectra in Figure 10:

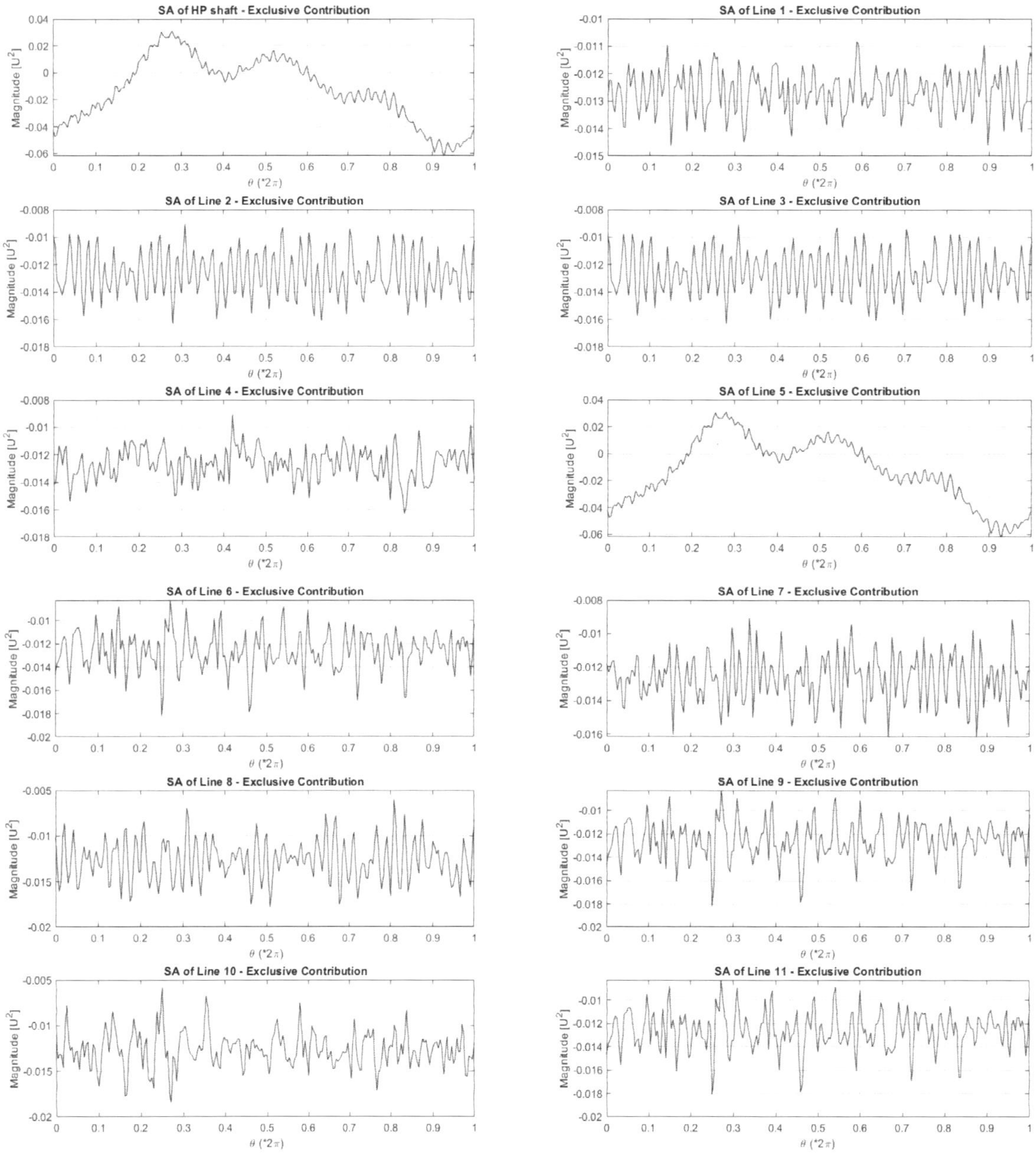

Figure 9. The signals of the 12 sources extracted from the AGB of the CFM56

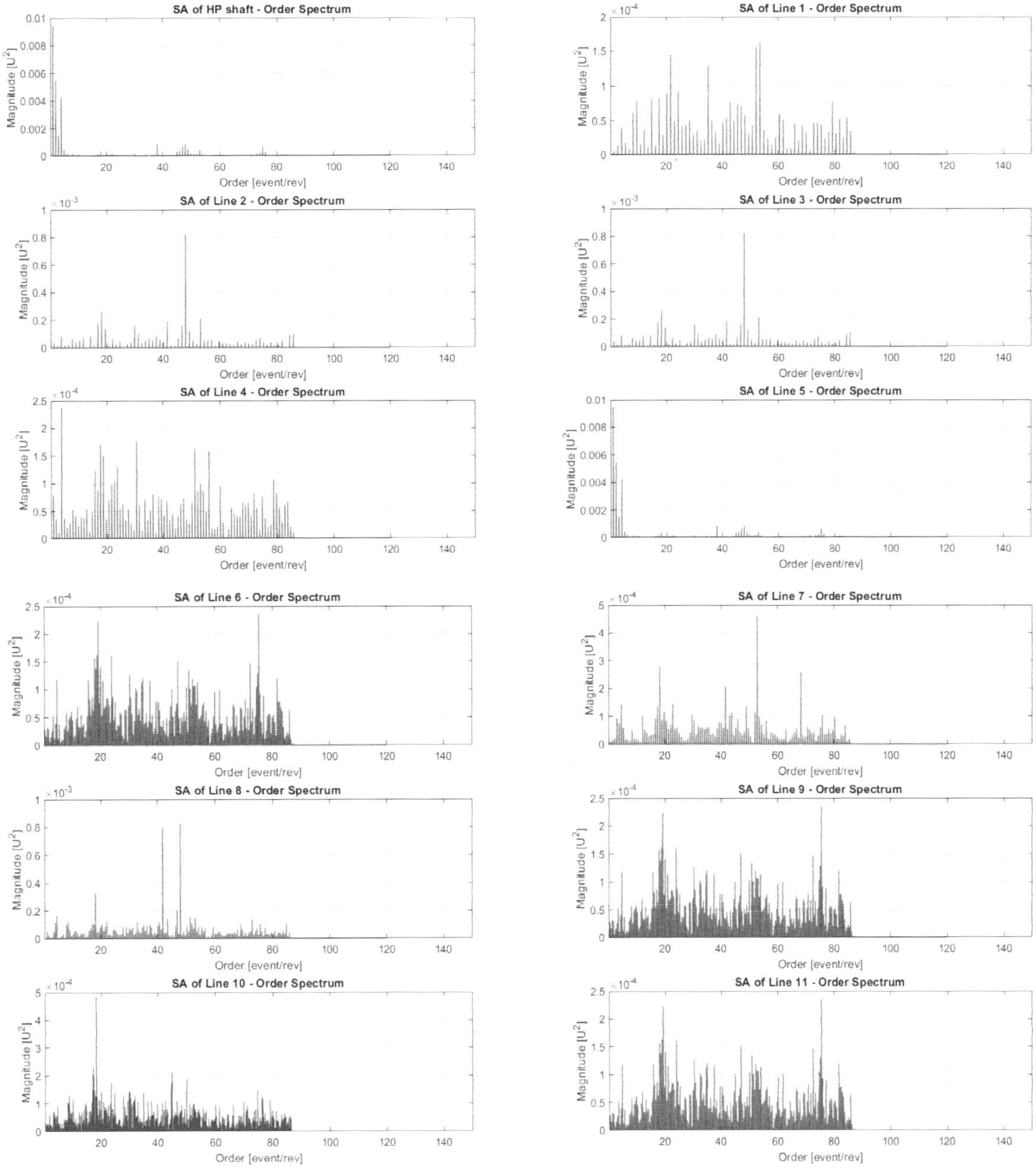

Figure 10. The order spectra of the 12 lines

The proposed methodology successfully extracts the AGB's deterministic signal and its constituent lines while eliminating the interferences and its replication for subsystem contribution. However, as it is seen in the MCO (Figure 7) and in the extracted lines with their spectra (Figures 9-10), it is possible for some shafts to share the same order. To overcome this issue, multiple input sensors can be employed to identify the exclusive contribution of each shaft while using the same proposed methodology.

V. CONCLUSION:

This paper presents a new approach for isolating the subsystems in a complex mechanical system, specifically an aircraft accessory gearbox. The method leverages synchronous averaging, set theory, and the matrix of common orders to eliminate redundant frequencies and uncover the unique contributions of each subsystem. This allows for a cleaner analysis for having each contribution well isolated from interferences, as well as the ability to pinpoint specific issues within a subsystem. Overall, the proposed method can improve the maintenance and global performance of mechanical systems while providing a good understanding of the behavior of complex mechanical systems.

ACKNOWLEDGEMENTS:

F. Karkafi gratefully acknowledges the European Commission for supporting the MOIRA project (Grant Agreement 955681) through the H2020 Marie Skłodowska Curie Innovative Training Networks program.

REFERENCES:

[1] R. B. Randall and J. Antoni, "Rolling element bearing diagnostics—A tutorial", Mech. Syst. Signal Process., vol. 25, no. 2, pp. 485–520, 2011.
[2] S. Braun, "Mechanical signature analysis: theory and applications", 1986.
[3] J. Antoni and R. B. Randall, "Differential diagnosis of gear and bearing faults", J Vib Acoust, vol. 124, no. 2, pp. 165–171, 2002.
[4] J. Antoni, F. Bonnardot, A. Raad, M. El Badaoui, Cyclostationary modelling of rotating machine vibration signals, Mechanical Systems and Signal Processing 18 1285–1314, 2004.
[5] J. Antoni, "Cyclostationarity by examples", Mechanical Systems and Signal Processing, 23 987–1036, 2009.
[6] J. Dion, C. Stephan, G. Chevallier, H. Festjens, "Tracking and removing modulated sinusoidal components: A solution based on the kurtosis and the Extended Kalman Filter", Mechanical Systems and Signal Processing, 38 (2), pp.428-439, 2013.
[7] M. Bai, J. Huang, M. Hong, F. Su, "Fault diagnosis of rotating machinery using an intelligent order tracking system", J. Sound Vib. 280, 699–718, 2005.
[8] J.R. Blough, Development and analysis of time variant discrete Fourier transform order tracking, Mechanical Systems and Signal Processing 17, 1185–1199, 2003.
[9] Y. Guo, K.K. Tan, "Order-crossing removal in Gabor order tracking by independent component analysis", J. Sound Vib. 325, 471–488, 2009.
[10] M.-C. Pan, C.-C. Chiu, "Investigation on improved Gabor order tracking technique and its applications", J. Sound Vib. 295, 810–826, 2006.
[11] M.-C. Pan, S.-W. Liao, C.-C. Chiu," Improvement on Gabor order tracking and objective comparison with Vold–Kalman filtering order tracking", Mechanical Systems and Signal Processing 21, 653–667, 2007.
[12] H. Vold, J. Leuridan, "High Resolution Order Tracking at Extreme Slew Rates, Using Kalman Tracking Filters", SAE Paper No. 931288, 1993.
[13] H. Vold, H. Herlufsen, "Multi axle order tracking with the Vold–Kalman tracking filter", Sound Vib. 31 (5), 30–34, 1997.

[14] Y. Guo, K.K. Tan, "High efficient crossing-order decoupling in Vold–Kalman filtering order tracking based on independent component analysis", Mechanical Systems and Signal Processing 24, 1756–1766, 2010.
[15] M.-C. Pan, C.-X. Wu, "Adaptive Vold-Kalman filtering order tracking", Mechanical Systems and Signal Processing 21, 2957–2969, 2007.
[16] K.S. Wang, P.S. Heyns, "The combined use of order tracking techniques for enhanced Fourier analysis of order components", Mechanical Systems and Signal Processing 25, 803–811, 2011.
[17] P. D. McFadden, "Interpolation techniques for time domain averaging of gear vibration", Mechanical Systems and Signal Processing vol. 3, no. 1, pp. 87–97, 1989.
[18] F. Bonnardot, M. El Badaoui, R. B. Randall, J. Daniere, and F. Guillet, "Use of the acceleration signal of a gearbox in order to perform angular resampling (with limited speed fluctuation)", Mechanical Systems and Signal Processing vol. 19, no. 4, pp. 766–785, 2005.
[19] S. Braun, "The extraction of periodic waveforms by time domain averaging", Acta Acust. United Acust., vol. 32, no. 2, pp. 69–77, 1975.
[20] P. D. McFadden, "A revised model for the extraction of periodic waveforms by time domain averaging", Mechanical Systems and Signal Processing vol. 1, no. 1, pp. 83–95, 1987.
[21] E. Bechhoefer, M. Kingsley, "A Review of Time Synchronous Average Algorithms", Annual Conference of the Prognostics and Health Management Society, 2009.
[22] J. Antoni, J. Griffaton, H. André, L. David Avendaño-Valencia, F. Bonnardot, O. Cardona-Morales, G. Castellanos-Dominguez, A. Paolo Daga, Q. Leclère, C. Molina Vicuña, D. Quezada Acuña, A. Partogi Ompusunggu, E. F. Sierra-Alonso, "Feedback on the Surveillance 8 challenge: Vibration-based diagnosis of a Safran aircraft engine" Mechanical Systems and Signal Processing 97, 112–144. 2017.

Where Theory meets Practice

ASSET HEALTH MANAGEMENT UTILIZING BATCH MULTIVARIATE PATTERN ANALYSIS

Chance M. Kleineke and Michael T. Santucci
Engineering Consultants Group, Inc.
3394 W. Market Street
Akron, OH 44333
330.869.9949
kleinekec@ecg-inc.com

Chance Kleineke, Systems Engineer, Engineering Consultants Group, Inc. Chance has a B.S. in Biomedical Engineering – Biomechanics from the University of Akron. He has worked for ECG since. He is a Project Manager for customer predictive analytics projects implementing ECG solutions. Involved with research, design, and development for software technologies. Primarily deals with software integration as well as maintenance, reliability, and performance optimization of various plant assets.

Michael Santucci, President, Engineering Consultants Group, Inc. Founder and president of ECG, Mike also acts as Chief Technology Officer for asset analytical solutions. He has over 30 years of experience in the Power Generation sector and is an expert in power plant operations, reliability excellence, predictive maintenance strategies, digitization, monitoring, as well as asset health and performance modeling.

ABSTRACT:

Budgetary and performance demands have led many power utility businesses to focus on some form of centralized fleet monitoring. Advanced pattern recognition (APR) is a common tool used for monitoring the major assets involved in the production of electricity in this manner. While pattern recognition techniques can focus on real-time steady-state operation and ease the challenge of monitoring hundreds or thousands of pieces of equipment, many assets can incur damage during the start-up and shutdown transient conditions that are not as commonly watched. This damage leads to failure over time, but signs of the damage may not be apparent during steady-state operation.

A multivariate pattern analysis was designed to identify anomalies specific to start-up and shutdown data of these assets, along with any other batch production process with defined start and end parameters. This method can be used across all types of processes and equipment. An early warning case study was conducted with a major power utility to validate the technique on a forced outage caused by a steam turbine bearing failure. Data was provided for two units with multiple coast downs over several months. This case study examines normal operation for monitored variables, such as temperatures and vibration readings, with the goal of detecting an individual variable's deviation from acceptable conditions and identifying the overall impact each has on the process. The study shows how this technique may be used to increase the early detection of faults and can be used in conjunction with other APR and diagnostic tools to help a business improve overall asset health management.

KEY WORDS: Advanced Pattern Recognition; asset condition diagnostics; centralized monitoring; condition-based maintenance; fleet-wide monitoring; health management; Multivariate Data Analysis; predictive maintenance; startup and shutdown analysis; transient processes.

INTRODUCTION:

Advanced Pattern Recognition, or APR, is a technique used across many industries to manage substantial amounts of business data, whether it be for workflow, performance, maintenance planning, or other key performance indicators. Energy utilities commonly utilize some form of APR to monitor key assets of operation. This is because of the ability to scale monitoring from single assets, to units, to a fleet of units, easily.

It is becoming commonplace to see utilities today setting up monitoring centers to watch over multiple sites within a company's purview to centralize the maintenance and performance optimization roles. This not only consolidates the required manpower for observation, but it allows for the sharing of information and failure mode analysis across all the assets within a fleet. A concern for monitoring centers is how to best manage the thousands of asset variables that will be continually analyzed every day.

Companies continue to make advancements in maintenance practices to control costs and improve equipment uptime. In the past, utilities would operate on a run to failure mentality, only performing maintenance on equipment when it failed or had easily noticeable symptoms, i.e., hot to the touch, audible noises, etc. Companies then turned to preventative maintenance techniques such as vibration and oil analysis to indicate pending issues with equipment. Many have now moved into predictive maintenance practices, as more instrumentation is used to monitor numerous variables of a machine. This added data can be used to plan maintenance in advance to reduce downtime and costs associated with any lost production. This increases the opportunity window to repair the equipment, Figure 1. APR is a commonly used tool in predictive maintenance programs today.

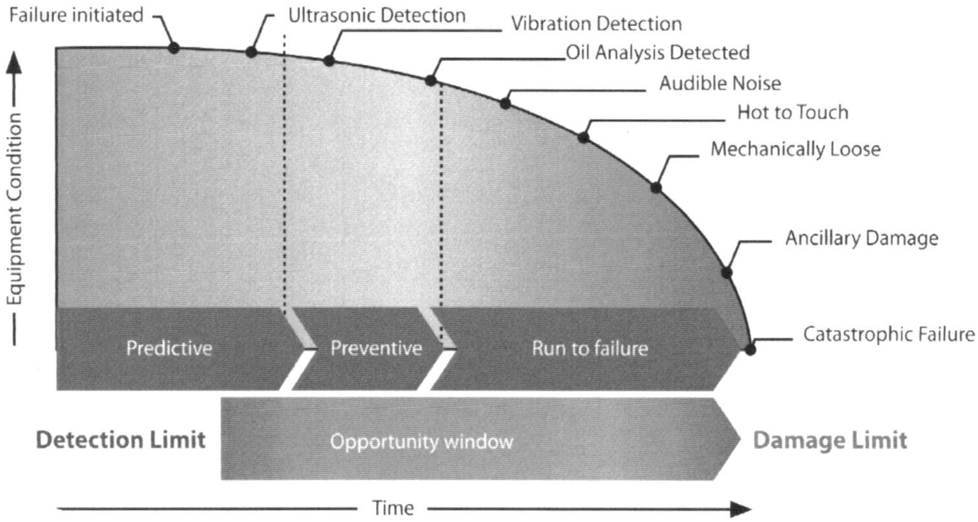

Figure 1. P-F Curve for failure detection.

APR has many advantages over standard hard-limit monitoring [1] where an alert is provided to analysts only when a particular variable extends outside of a set "optimal" range, Figure 2. This "optimal" range must be set to include all operating loads and external input to avoid false alarms. It does not consider factors such as operating load, ambient temperature, etc. APR methods rely on historical data to predict the value of each asset variable at a given moment. The predictions are based on the activity of all other variables being used in a particular model for that machine, and it can account for externalities and load.

Figure 2. Early warning detection using APR modeling.

These techniques have been proven effective at early warning fault detection on a large scale. But APR methods typically focus on steady-state operation. When models are built, they are commonly only running against the data once the equipment reaches a particular load/state. The models are shut off during transient states and when the equipment is offline. This is because it is harder to use standard APR techniques to "learn" the behavior of the equipment while operating states are changing quickly or to a large degree. Transient conditions contain a great deal of information but are typically not well monitored.

PATTERN RECOGNITION DURING TRANSIENT CONDITIONS:

A transient process has a start and end time, as compared to a continuous process that can last for any given time. During a defined transient state for an asset, repeated behavior can be expected for monitored variables. The process is dynamic and can have varying time constants, as the equipment may not have reached steady state operation. This can be a short duration event, such as a startup or shutdown of a gas turbine.

Goals of incorporating transient conditions into the pattern recognition package included enabling easy startup and shutdown interpolation and providing actionable insights to alert analysts when a startup needs further review. This will allow faster troubleshooting for problematic startups and aid in the root cause analysis of failures. Closely monitoring the startup and shutdown states of assets will reduce the risk of equipment failure and maximize runtime. Typically, APR models focusing on steady-state operations will be shut off during these transients in addition to when the asset is offline. Figure 3 shows differing benefits of traditional APR and transient monitoring.

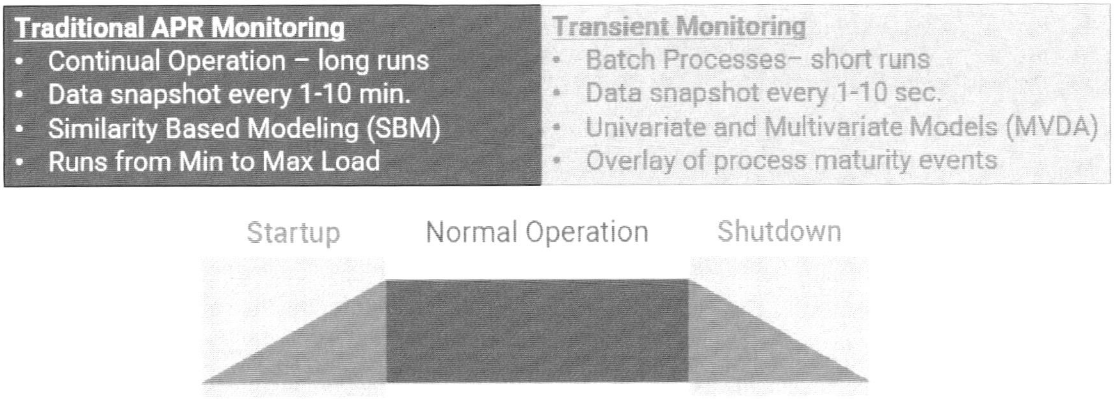

Figure 3. Traditional APR Monitoring vs Transient Monitoring.

A set of asset variables must be considered for the process, typically 20 to 100. The method used is a modified version of modeling batch processes [2] in a 3-D format considering the number of variables in a batch, the time intervals each batch, and the number of normal batches, variables *x* time *x* starts, Figure 4. The method deploys both univariate SPC and Multivariate Data Analysis (MVDA). MVDA is a statistical technique used to analyze data that is generated from more than one variable. An MVDA model is a summary of individual variables statistically organized together using a specific algorithmic function.

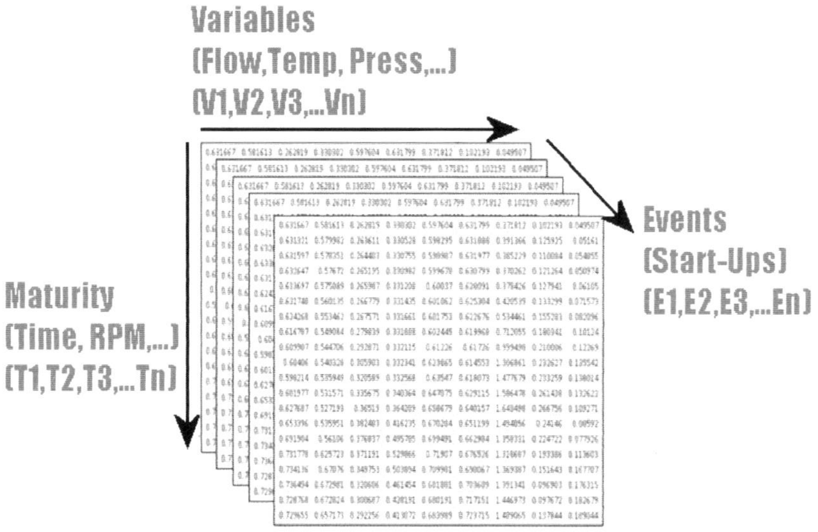

Figure 4. Modeling batch processes with a 3-dimensional matrix.

A monotonic maturity variable is used to define start and end times of a batch or transient Time is a common maturity variable, or in the case study to follow, RPM is used. A set of "golden" batches are selected to represent the boundaries of normal/optimal operating conditions. A series of single variable plots is used to view across multiple events. Multidimensional reduction identifies primary features of a model, then reduces further to characteristics of the signal and the noise [3]. This focuses the multiple variables in a model into two core features defining the process. The signal data is used to detect strong deviations in systemic parts of the data, while the noise limit detects deviations from the overall correlation structure, a residual analysis similar to standard APR.

For model creation, the process variables to be monitored must be determined, i.e., temperatures, vibration data, pressures, flows. Then the maturity variable must be defined, i.e., time, speed. Next, a discovery range is set, defining which events to view as optimal. The event can then be defined with parameters defining the start and end of the transient state. These factors can then be used to find all batch training events for building the model, Figure 5.

The data for training will be pulled in at a set interval and overlaid for each variable. Once the data is retrieved, bad data and outlier batches can be removed. This would include stale data, temporary spikes, bad batches, etc. The model can then go live against current data, analyzing any new startups that occur.

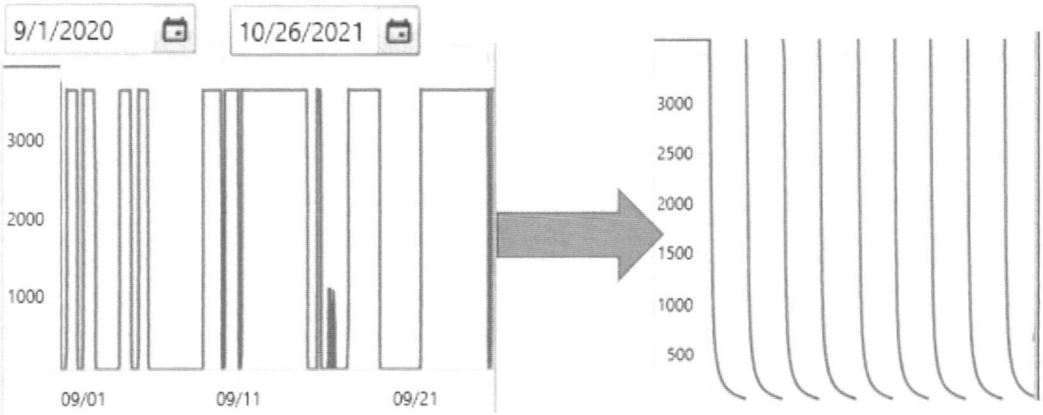

Figure 5. Isolating transient training conditions from the discovery range.

STEAM TURBINE COASTDOWN STUDY:

A case study was performed on a steam turbine that had a bearing failure in late September 2021. Data was provided for two units with multiple coastdowns over several months, the failure was on Unit 1. This case study examines normal operation for monitored variables, such as temperatures and vibration readings, with the goal of detecting an individual variable's deviation from acceptable conditions and identifying the overall impact each has on the process. The M&D center had no APR alarms prior to the failure. The transient pattern recognition method was applied to see if it could forecast the failure ahead of time focusing on turbine coastdowns. A coastdown refers to the process of removing a turbine from use, as it decreases from full load to shutting off. The bearings of a turbine encounter considerable loading forces during coastdown, and so should be monitored closely for potential problems. For the study, two years of coastdown data was provided. During this time there were 48 coastdowns with 71 variables captured for the steam turbine that had a failure.

Coastdowns characterized as normal were used as the baseline. Statistical process control limits were calculated for each variable at every speed during the coastdown. An MVDA model was created to represent the similarity of each coastdown and the projection of each variable. Anomalies can then be detected either by SPC limits or model irregularity. The turbine speed was used as the maturity variable, looking for an event where the speed fell between 50 and 3590 RPM (operating speed is 3600 RPM).

All events within the discovery range can be overlaid and a clear pattern is easily recognized, Figure 6. Two batches were discarded due to accelerated time to shutdown. The remaining coastdowns are used to calculate univariate alarm thresholds.

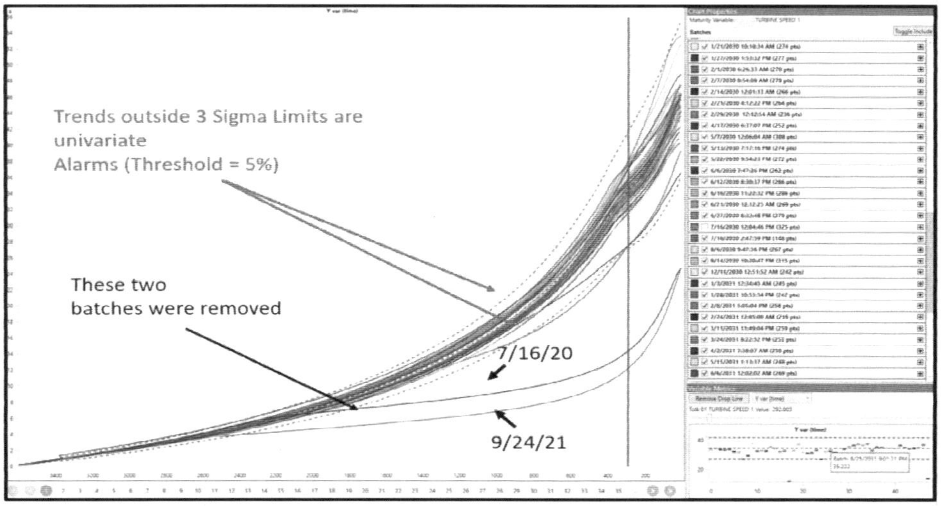

Figure 6. Determining outlier batches.

Variables used in the model include turbine bearings 1-9 metal temperatures, bearings 1-9 oil drain temperatures, bearings 1-9 rotor vibrations, turbine rotor positions, and others. Looking at all variables within the model, anomalies in February of 2021 show up for turbine metal temperatures and vibrations, Figure 7.

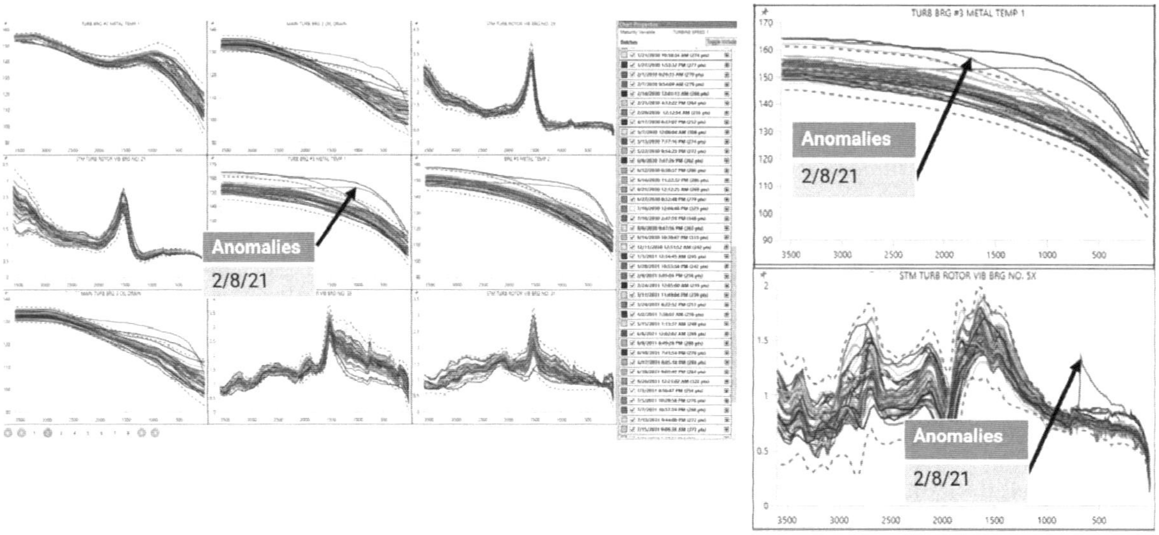

Figure 7. Anomalies within the model at an individual variable level.

Several coastdowns show temperature spikes on bearings 4 or 5 starting in December 2020, Figure 8 and 9. This temperature spike consistently occurred late in the coastdown, typically as the turbine decreased from 550 to 200 RPM. A sustained step change in the bearing 4 and 5 metal temperatures would have triggered an alarm after Jan 2021, giving 9 months early warning prior to the failure. A sustained step change in the bearing 6 metal temperatures in Figure 10 also would have triggered an alarm after June 2021.

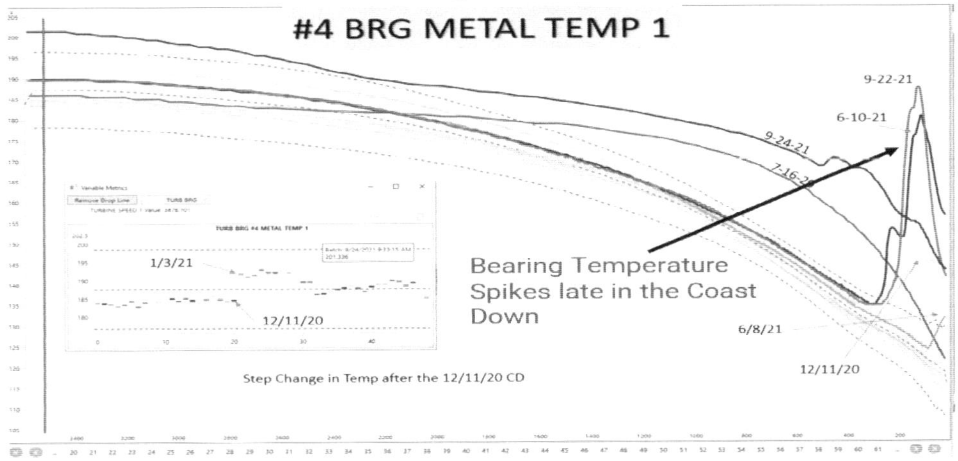

Figure 8. Unit 1 Bearing #4 metal temperature showing step change in data.

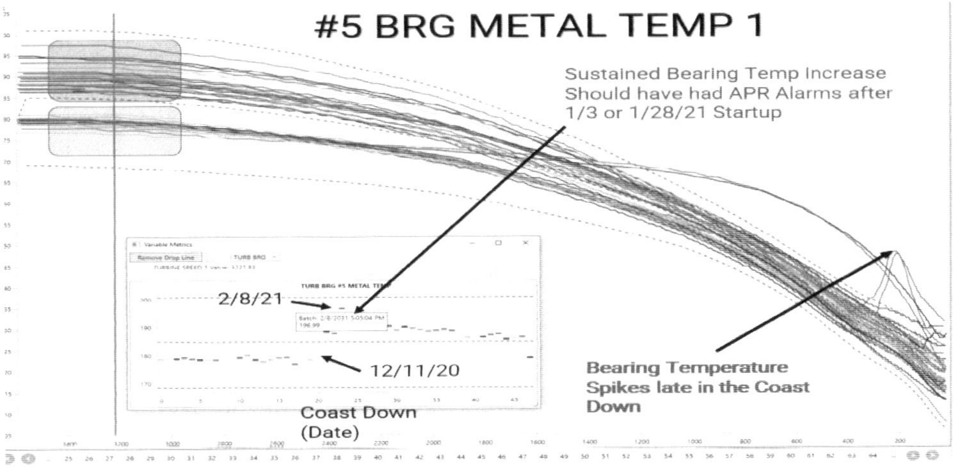

Figure 9. Unit 1 Bearing #5 metal temperature showing step change in data.

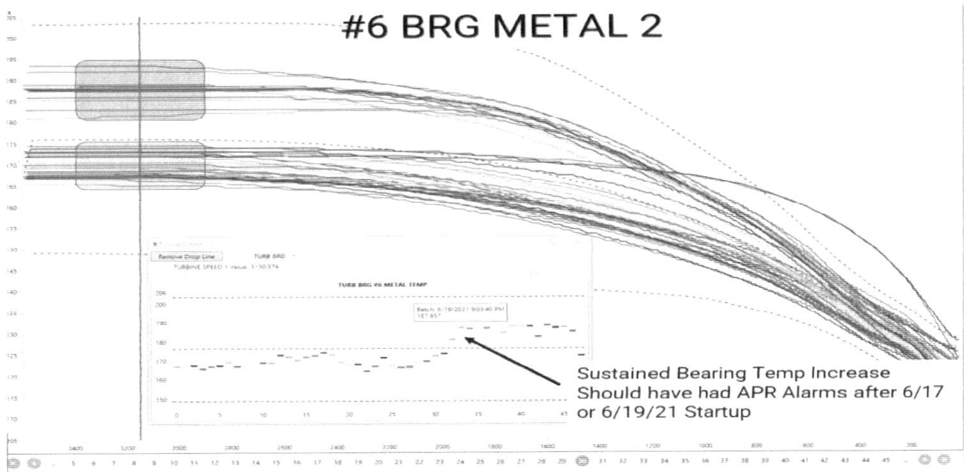

Figure 10. Unit 1 Bearing #6 metal temperature showing step change in data.

RESULTS:

The Unit 1 forced outage occurred in September 2021. After the failure, failure analysis found indications of bearing damage primarily on bearing 4. Bearings 4, 5 and 6 had considerable damage and were rebabbited. The root cause of the damage was believed to be due to the operation of the lift oil pump during unit coastdown. The lift oil pump, Figure 11, turns on at lower RPM to circulate oil when the momentum in the turbine is too low to properly distribute oil. All lift oil pumps were replaced on this unit. As a result of this study, the turbine speed for the pumps to come online during coastdown was changed from 200 RPM to 600 RPM, as the temperature would increase before the turbine got down to the original setting of 200 RPM. The pump pressures are now monitored when the turbine rotor goes on turning gear.

Figure 11. Lift oil pump for a steam turbine.

Unit 2 data was simultaneously studied. While conducting the failure analysis on Unit 1, this method found a parallel developing issue on Unit 2 bearings. The customer scheduled an outage on Unit 2 to inspect the main bearings. Bearings of concern from the study were 5 and 6. Bearings 3, 4, 5 and 6 were all rebabbited. Similarly, to Unit 1, all lift oil pumps were replaced for Unit 2 and turbine speed for the pumps to come online was adjusted, and the pump pressures are now monitored when the turbine rotor goes on turning gear. Figures 12 and 13 show resulting charts on a single bearing metal temperature for each of the units studied. There were 23 coastdowns on Unit 1 following the first observation of increased bearing temperature leading to the turbine failure. There were 12 coastdowns on Unit 2 following the first observation in the study, and maintenance was performed to prevent catastrophic failure. Figure 14 shows the damage to two of the bearings.

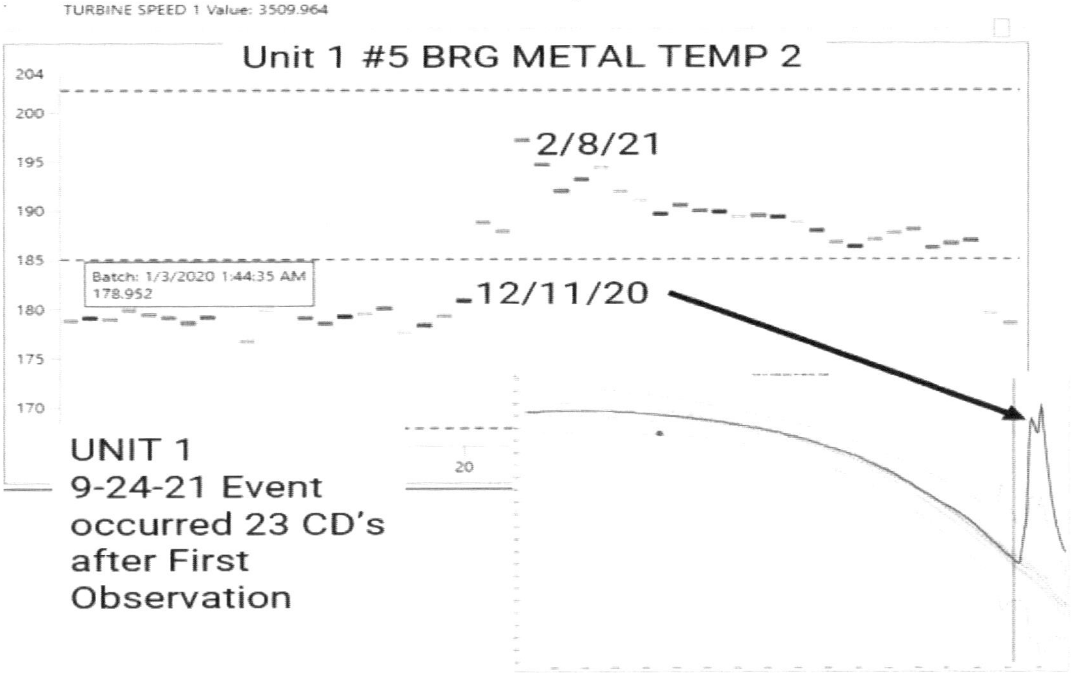

Figure 12. Results for Unit 1 Bearing #5 temperature increase leading up to failure.

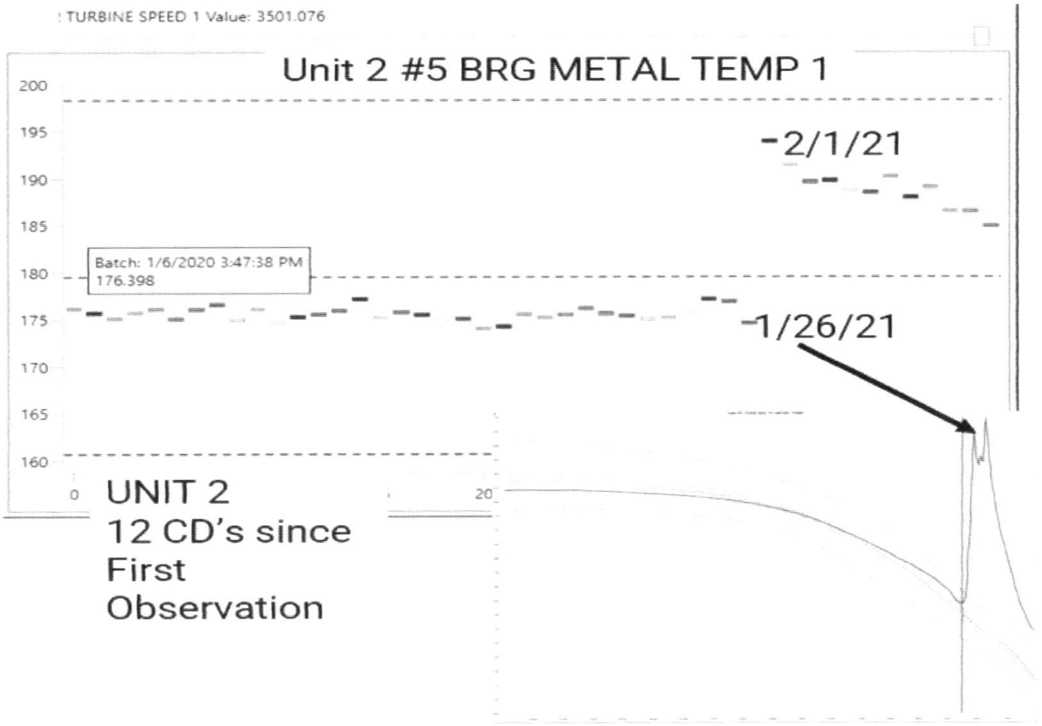

Figure 13. Results for Unit 2 Bearing #5 temperature increase prior to maintenance.

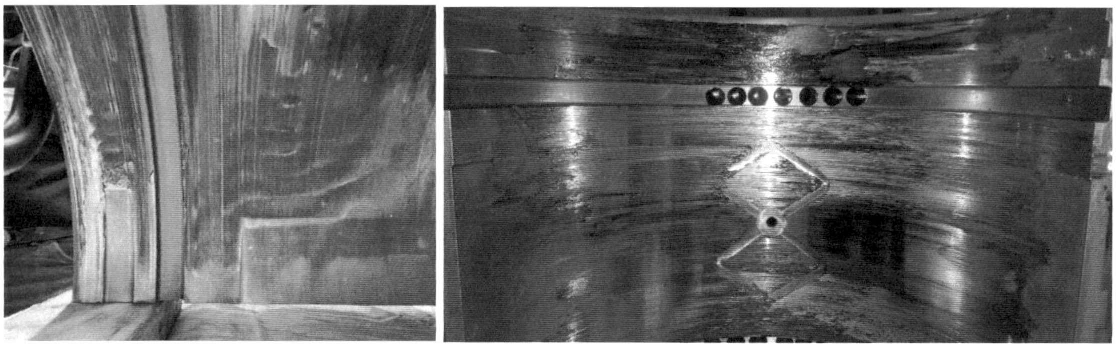
Figure 14. Bearing race shows damage from lack of proper lubrication.

CONCLUSION:

The root cause was believed to be a combination of multiple start/stops of the shaft and the operating pressure of the lift pumps. In the past, there were fewer shutdowns of the turbine, but as other sources of electrical generation have become available, the need for these turbines has dropped and requires more cycling of the units. With fewer shutdowns, the turbine speed that would activate the lift pumps was never adjusted. The plant has indicated that they have gone from roughly a dozen start/stops per year to multiple start/stops within a week sometimes, easily approaching 100 or more cycles in a year, including long time periods with the turbine on turning gear to keep the unit on hot standby. This has led to wear and increased clearances within the turbine.

The speed at which the pumps turn on was increased to start the lift oil pumps sooner in the process. The pressure was increased with the new pumps, and how the plant monitored the pump pressures was changed as a result of this study. The plant now monitors start/stops of the turbine very closely with a dedicated observer in the control room watching critical parameters displayed graphically.

REFERENCES:

[1] Kleineke, C., Misal N. and Santucci M., 2010, An Application of Pattern Anomaly Detection Methods to Fleet-Wide Asset Level Diagnostics, *MFPT Proceedings 2018*

[2] Wold, S., Kettaneh, N., Friden, H. and Holmberg, A., 1997, Modelling and diagnostics of batch processes and analogous kinetic experiments, *Chemometrics and Intelligent Laboratory Systems 44 (1998) 331-340*

[3] Westerhuis, J., Gurden, S. and Smilde, A., 2000, Generalized contribution plots in multivariate statistical process monitoring, *Chemometrics and Intelligent Laboratory Systems 51 (2000) 95-114*

WHAT DRIVES ACCURACY FOR VIDEO VIBRATION?

Bill Marscher, Andy Lerche, and Sergey Frolov
Mechanical Solutions, Inc.
11 Apollo Drive
Whippany, NJ 07981
(001) 973-326-9920
AHL@mechsol.com

Abstract: Vibration detection and display using magnification, or amplification, of high-speed video (Vibration Video Amplification, VVA) has been demonstrated to be a very useful tool. VVA has been challenged to produce useful results, especially when lower quality cameras are used, or if the component being tested involves relatively high frequencies (e.g. 200 Hz or higher). For these cases, detection of very low displacement levels is required for VVA to give useful results. This is because video methods detect displacement, not velocity, and vibration standards, such as ISO 20816 or HI 9.6.4, are based on velocity. For a given velocity acceptance level, the displacement associated with that velocity decreases in proportion to the vibration frequency. At relatively low frequencies such as 50 or 60 Hz, the resulting displacements are easily detectable. However, at higher frequencies such as gas turbine gas generator ("high spool") operating speeds, or pump or compressor blade passing frequencies, it is much more challenging. At 500 Hz for example, the detection capability required for vibration detection is roughly equal in terms of mils versus in/sec rms. Therefore, to detect 0.2 in/sec rms (5.1 mm/s), displacement detection accuracy of 0.2 mil p-p (about 5 microns p-p) is required. This is achievable, but challenging for most cameras and scene evaluation software. Therefore, this paper will discuss methods to improve and evaluate the displacement resolution of VVA. The encouraging degree of precision that has been able to be achieved is reported.

Key Words: Motion Magnification; Measurement Accuracy; Rotating Machinery; Machinery Diagnostics; Vibration Video Amplification

1. INTRODUCTION

Turbomachinery, pump, and motor vibration is often useful in determining whether or not a machine is operating properly, and for diagnosis of problems if the operation appears improper, or if reliability issues have been experienced (e.g. fatigue cracking, or premature failure of bearings and seals). For several decades, visual methods have animated a CAD model to exhibit operating deflection shapes (ODS) of vibration in an understandable manner. This has been an important tool in getting a complete and simultaneous view of the vibration of the machine and its system (e.g. piping, foundation, and driver or driven machine), so that the potential importance of any given vibration mode is placed in proper context. The resulting animations represent the motion implied by phase-linked signal spectra from the group of accelerometers. To provide representative and comprehensive results, such accelerometers typically need to be located over hundreds of test points. Then the exaggerated (but to-scale) vibrations can be animated as "cartoons" at frequencies where the response is sufficient to be of interest to the troubleshooter. This process may take one or more days for complex machinery and systems.

An innovative vibration detection and display method (VVA) has been developed by several independent groups. Each group's approach has a different algorithm at its core, but all are based on evaluation of high resolution/ high speed video taken of the vibrating component, such as operating machinery. These VVA methods can provide information equivalent to a high-sensor-count ODS test. Several methods, including that of the authors, does this by treating each pixel in a video scene as an accelerometer using the pixel's (or group of pixels') light intensity modulation statistics to determine the local vibration displacement motion. From this information, realistic magnification of slow-motion video footage permits microscopic vibration to be magnified and thereby observed and interpreted by the human troubleshooter. Just as with accelerometer outputs, the pixel statistics can be Fourier transformed to determine frequency spectra as well. The VVA-based vibration display method is much faster than classical ODS, yet provides similar information at many more locations (at least in directions within the 2-D field of view), with less opportunity for human error, and in a manner that is more credible to both technical as well as non-technical decision makers because they are looking at the actual machine or structure. However, use of VVA at high frequencies, or in other situations where even excessive vibration displacements (according to standards such as ISO) are very low, has been an ongoing challenge for commercially available VVA systems. This is particularly true when lower cost camera equipment is implemented.

2. Background

Classical Vibration Visualization Methods:

About 1980, some researchers developed a methodology for "seeing" vibration in extensive mechanical systems. Their approach was to acquire data at many locations and directions on a vibrating structure, and then allocating those motions to a stick-figure CAD model. This model was then animated on a computer video screen, or the extreme motion was printed in a still frame. The resulting ODS method has proven very powerful, but is time-consuming, and therefore expensive, to implement in any detail. It is also prone to user error, with regard to book-keeping the sensor locations and directions, and accurate construction of the cartoon model.

Vibration Video Amplification (VVA) Methods:

Many of the drawbacks can be avoided if an actual visual scene is able to be evaluated concerning the vibration of components within the scene. Academic researchers, some of whom are listed in the References [1], [2], and [3], had this thought as much as 30 years ago (albeit not for the purpose of modal testing at the time), and have been gradually perfecting techniques. Basically, these techniques fall into two categories: 1) tracking of specific points, edges, or (as machinc-vision scientists call them) "blobs", and 2) performing statistics, including signal vs. time as well as signal FFT frequency spectra, on the individual independent pixels. The former are called "Lagrangian Methods" by many researchers, and the latter are called "Eulerian Methods". A rich literature basis exists for these methods. This paper's limited references emphasize the Eulerian technique.

The Vibration Video Amplification (VVA) methods may be summarized as follows:

- Uses high-speed, high-resolution video
  - Eulerian method is equivalent of millions of accelerometers, 1 per pixel
- Analyzes/quantifies displacement motion, in 2-D
  - FFT can be used to separate motion into individual frequencies
- Algorithms amplify motion to human visual threshold
  - Filterable by desired frequencies, high resolution (up to 4+ kHz in the system as presented by the authors).
  - Amplifies up to 1000+ times the actual motion, using a statistical algorithm that can perceive and magnify motions down to less than 0.1% of pixel.

History:

Various researchers, in the US as well as Europe, have developed a variety of video acquisition and evaluation processes to study other dynamic phenomena. In fluid dynamics, laser-doppler velocimetry (LDV) has been able to track streamlines, and similar processes have been applied to mechanical vibration. In a lower-tech method, strobe lights have been used on machinery for many decades to "freeze" motion. These methods have been useful for interpreting dynamic behavior if relatively large motions are involved. However, displacement magnification was not achieved.

In the 1920's, advanced research organizations used relatively high-speed cinematography to guide development and to evaluate the validity of new vibration analysis procedures. A good example is the work of Wilfred Campbell at GE. At the time, steam turbines driving generators were growing larger, and as they did, they encountered unexplained fatigue failures. Campbell's work, as exemplified by the photo in Figure 1 (a single frame of a "movie" or video taken by his cameras), determined that bladed disks had much more complicated natural frequencies and resonant vibration behavior than had been predicted up to that time, including zero-vibration nodal lines in their mode shapes, now known as nodal diameters and nodal circles. When the associated natural frequencies matched a strong excitation frequency (such as the number of stator nozzles times running speed), and when simultaneously the mode shape of the natural frequency matched the lobes of a circumferential static pressure pattern (such as associated with the number of blades versus number of stator nozzles), a strong resonant response was seen to occur. This caused fatigue in many cases.

Based on this insight, GE and eventually others were able to avoid damaging resonances by either adjusting natural frequencies, or by adjusting problematic lobe patterns of the nozzle pass frequency. The Campbell Diagram, well known to turbomachinery engineers, became a graphical method of codifying this procedure.

Figure 1: Campbell's bladed disk video from 1924

Such successful efforts led scientists and engineers to an appreciation for the power of video to uncover useful information in dynamic scenes, assuming that appropriate algorithms, implemented by software, were implemented to evaluate the video footage.

Several universities, including MIT [1], [2] as discussed, have been actively pursuing VVA. At MIT, the particular technology that became the core of their research was an Eulerian approach [3], which tracks the variation of individual pixels over time, and then exaggerates any changes. To the human eye, no matter how long and hard you stare at a machine with vibration at, say, several times the ISO limits, a person would struggle to detect any motion. For a computer, however, the tiniest per-pixel fluctuations (between white and slightly-off-white, say) are easy to detect and quantify. MIT originally developed their software approach, known as Motion Magnification, to measure the vital signs of neonatal babies without physical contact, but they realized that there were other far-ranging applications. For biological applications, Eulerian VVA detects changes in skin color, as well as exaggerates movements.

Research by the Authors:

The authors' organization initiated its detailed research into engineering uses for high-speed video in 2003. The US Department of Defense contracted the authors' group to study how to make non-lethal projectiles operate reliably, ensuring that they functioned without harming humans or animals. Applications included various forms of "soft bullets". High speed video, up to 20,000 frames per second, documented their transient behavior during flight and impact.

A later application (2005) involved an inexpensive video sensor to use pattern recognition software to rapidly identify rocket-propelled grenades (RPGs) fired by a belligerent, so that defensive steps could be taken. Following this was use (2011 to present) of high-speed video to determine the moment-to-moment effectiveness of passive counter-measures to fool MANPADS missiles fired by adversaries to bring down (for example) an airliner, providing feedback facilitating improved countermeasure design. More recently (2015 to present), the authors' group was contracted to use high-speed video and pattern recognition to rapidly characterize lethal ranges of various munition types, so that friendly military forces are better able to ensure sufficient buffer zones for civilians, schools, and hospitals.

During this process, the authors began applying Eulerian algorithms to measure and amplify motions of individual pixels [4]. Different procedures were tried during both government and IRAD-funded testing, and after several years a successful method was achieved. Transient motions and vibration levels as low as two tenths of a mil (i.e. 5 microns) on machinery surfaces were demonstrated first in the laboratory, and soon after in on-site machinery troubleshooting. Recently, there have been further improvements in accuracy, as will be discussed.

As presented above, the authors' methods are based on the Eulerian approach [1], [2], [3], and [5], which observes the intensity variation of individual pixels, combining their effects statistically into a full scene. Alternative methods [6] typically keep track of motion pixel-to-pixel, the so-called Lagrangian techniques. Variations include feature tracking, and an approach called "optical flow". These Lagrangian methods are good for amplifying motions that occur over multiple pixels, e.g. motions of more than 2mm (0.080 inches) for a 3 m (10 foot) field-of-view. However, these approaches are limited in how small a displacement they can detect for a given field of view and a given camera resolution, in that vibration levels of interest can be well below such limits, particularly at higher frequencies of vibration. If higher camera resolutions are used to overcome this limitation, the computational overhead becomes excessive due to the mega pixel count. One competing application mediates this by selecting only a greatly reduced number of pixels in a scene, and using these as the basis for evaluation and animation.

The benefits of VVA performed for vibration testing are that it becomes a powerful and intuitive procedure for displaying complex patterns and vibrating shapes in a relatively short time-frame. It realistically animates modes of vibration in a manner that is easy for non-experts to believe and understand. Importantly for certain applications, the technique does not require contact, so that surfaces above 1000 F, or that are radioactive, or that are (for example) 50 feet vertically above in a ceiling, can be observed without difficulty. Furthermore, items such as machinery instrumentation wires or lubrication lines, which would change their behavior if mass-loaded with even a small accelerometer, can be evaluated in a manner that does not alter their behavior. The results are obtained quickly, and at modest cost and effort.

However, there are also several potential problems in applying VVA to vibration motion detection and display. The primary issue is whether displacement detection is sufficiently accurate and reliable. A short discussion of a math example illustrates this. Based on the physics and units conversion:

$$V \text{ in/sec rms} = 0.00222 \times \text{frequency} \times \text{Displacement mils p-p}$$

Applying this formula, for an example frequency of 500 Hz, 0.020 mils p-p, as demonstrated by the authors' research, translates into 0.022 ips rms, very adequate for comparison of vibration to standards such as ISO. However, 1 mils p-p (as is the typical limitation for some systems) translates into 1.1 ips rms, clearly not acceptable for detecting potentially serious vibrations, and is well in excess of standards. Figure 2 illustrates the displacement vs. frequency issue over a broad practical range of interest for machinery assessment, based on the ISO 10816-3 standard.

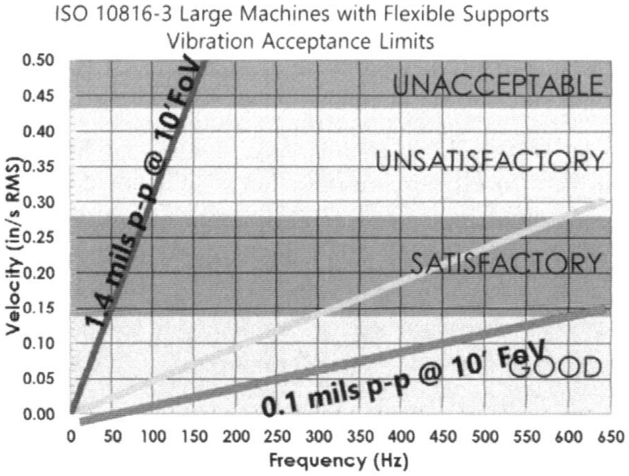

Figure 2: Relationship of Vibration Velocity vs. Displacement and Frequency [7]

Eulerian methods have been shown to work better than competing methods for detection of small displacements, and motion detection as low as 1/1000 of a pixel have been demonstrated [4]. When enhanced by triggered synchronous averaging, the authors have achieved results superior to this.

The concept of triggered synchronous averaging is an averaging-by-summing process that is performed for many data sets taken of a consistent steady state vibration, as represented by the output of various vibration detection sensors (usually accelerometers when VVA is not being implemented)

In the case of VVA, the authors have accomplished this by a proprietary approach involving multiple synchronized video clips being "averaged" together. This:

- Reduces level of vibration not coherent with the steady vibration
- Reduces camera sensor white noise effects
- Lowers vibration detection threshold for same duration video
- Enables use of lower cost camera equipment

Research Results:

The accuracy testing by the authors used the set-up shown in Figure 3:

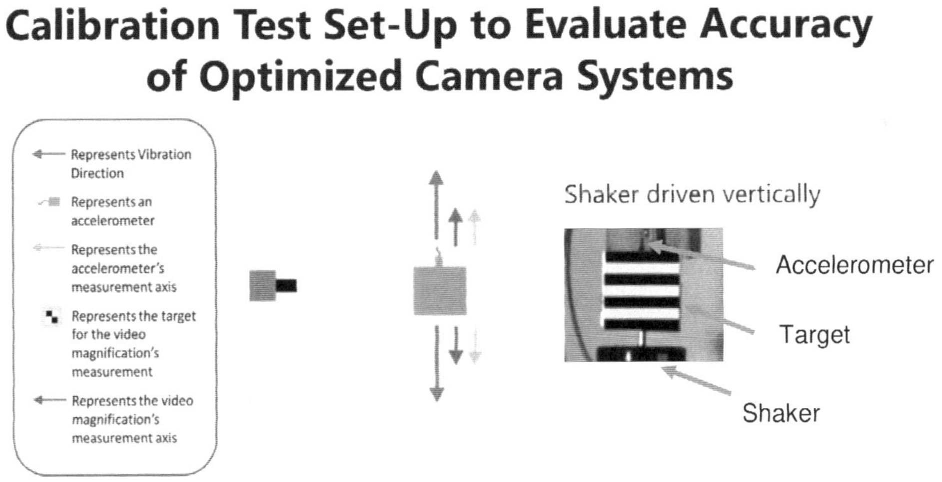

Figure 3: Accuracy calibration test set-up. Certain tests used proximity probes or a laser vibrometer in place of, or in addition to, the accelerometer.

The authors' research has shown that accuracy of detection during VVA depends strongly on the following issues:

1. **Hardware**
    - Low sensor noise required, need expensive cameras.
    - High bit depth of sensor ("dynamic range") required, again need expensive cameras.
    - Screen resolution: No longer such a big issue, inexpensive cameras OK.

2. **Acquisition**
    - Always a function of the field of view
    - Illumination level, contrast
    - Focus/edge definition, shutter control
    - Steadiness (shaky tripod)

3. **Software**
    - Nyquist requires at least a frame rate/frequency span factor of 2
    - Secret Sauces are different from vendor-to-vendor

Figure 4 through 7 show representative results from the authors' research, in all cases based on video taken using the triggering and time-averaging approach of the authors. Fig. 4 shows that, as displacement becomes lower, the error associated with the measurement goes down. This is a natural outcome of the error level having a "floor", that then acts as a numerator in the error percentage calculation. This becomes a particular problem below 0.1 mil p-p, if special approaches are not used, such as shutter control, and trigger timing and time-averaging of the video frame acquisition. With appropriate shutter control and frame acquisition and averaging, accuracy improves to order of 0.020 to 0.050 mil p-p, and error from the camera sensor noise floor also statistically becomes less of an issue.

Fig. 5 and 6 show that frame rate effects, and frequency effects as a percentage of frame rate, also have influence, with poorer accuracy able to be achieved at lower frame rates, or at higher frequency/frame rate ratios. In addition to the latter, higher frequency has an inherently large practical effect on amplitude accuracy for a given vibration velocity, as illustrated in Fig. 2. Fig. 7 shows that at low displacement amplitudes, accuracy is even more sensitive to frame rate, with low frame rate being a challenge for cameras constructed for high frequency video acquisition.

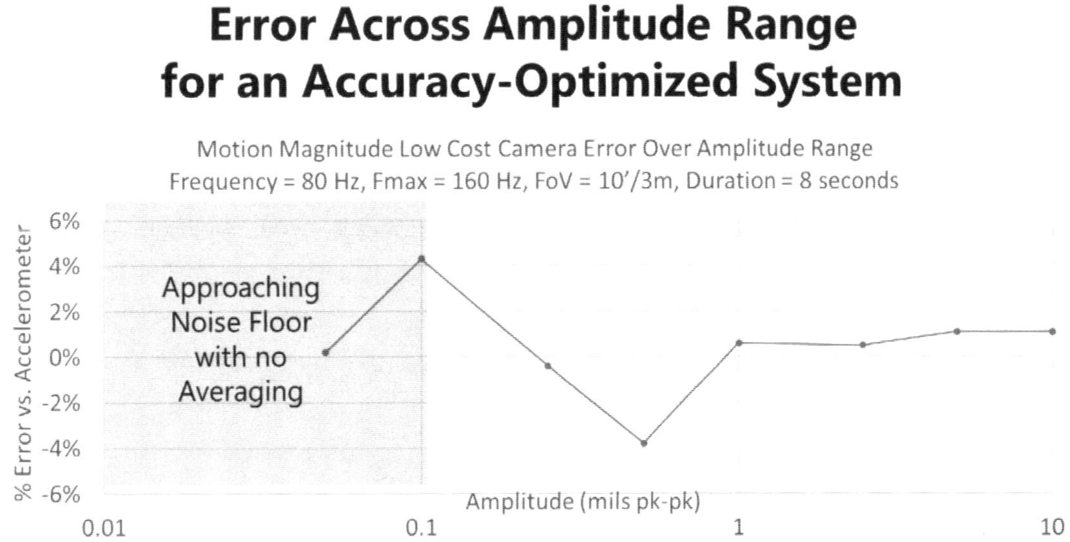

Figure 4: Accuracy test results showing error in detected vibration vs. amplitude

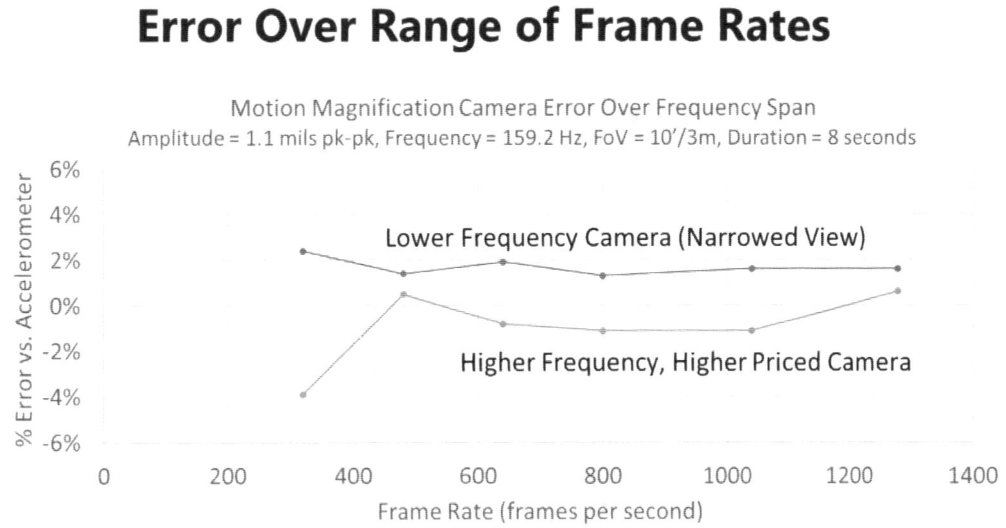

Figure 5: Accuracy test results showing error versus camera frame rate

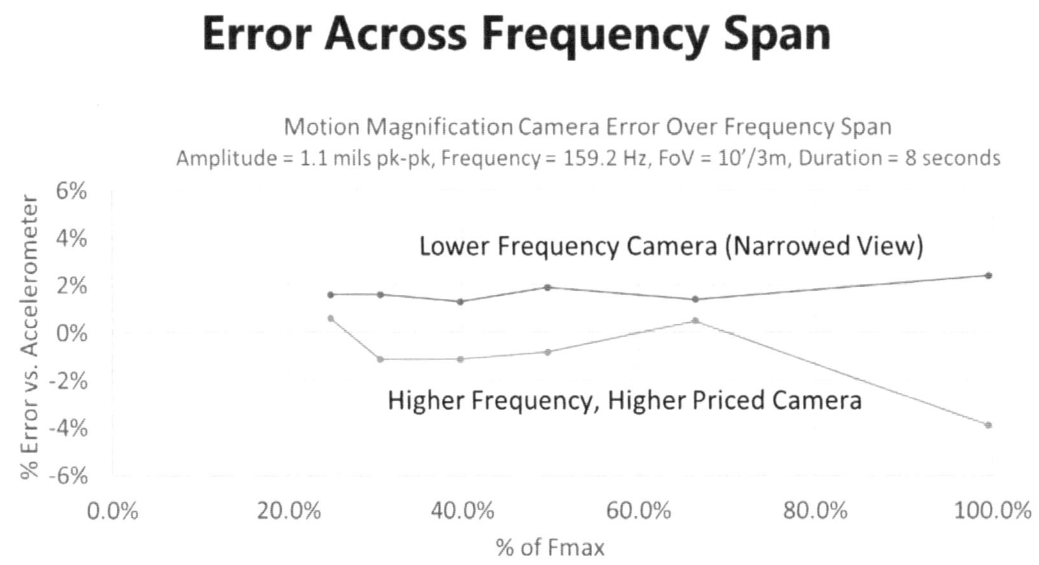

Figure 6: Accuracy test results showing error vs maximum frequency span (Fmax)

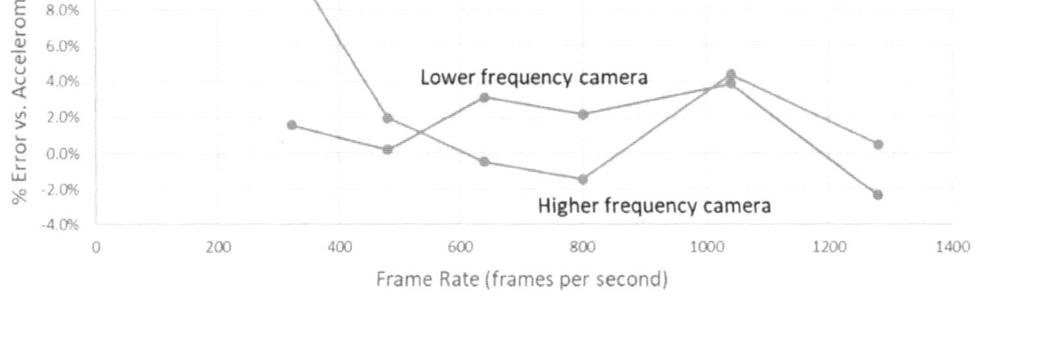

Figure 7: Accuracy test results showing error versus frame rate for lower vibration levels

A summary of the accuracy experienced by the authors during their research is as follows:

- With no accuracy optimization, author's experience was +/-35% error
- Recent research using advanced techniques achieved an accuracy within +/-5%
- Accuracy depends on the camera hardware, lighting, and algorithms used
- In fully optimized cases, +/- 2 to 4% accuracy can now be typical
- Best accuracy for higher frame rates
- Even <1% accuracy achievable consistently for 1 mil p-p vibration.

3. FUTURE RESEARCH

The following summarizes the research currently underway at the authors' organization:

1. Improving VVA capability using low-cost COTS cameras
2. Continuous improvement of displacement sensing accuracy
3. Continuous improvement of upper frequency limit
4. Acquiring ODS (including torsional) vibrational motion of rotating components
5. Processing and displaying 3-D vibration patterns

4. CONCLUSIONS

As presented, the new video-based vibration amplification procedures and equipment presently on the market can be very useful for determining and displaying vibration behavior of rotating machinery and associated piping and systems. The new method discussed in this paper has advantages over the conventional VVA, and is superior (or at least worthy to be supplemental) to traditional measurement methods in many instances. VVA methods typically take much less time and "logistics" than conventional detailed vibration testing, such as ODS methods, to implement.

Primary conclusions of this research include:

1. Detection accuracy of conventional video vibration analysis systems is often insufficient, if used to detect vibration levels relevant to standards, in a full field-of-view.
2. Detection accuracy is a complicated function of many factors, including (but not limited to) camera quality and video-taking technique.
3. Camera dynamic range and "noise floor" is also key (this is poor in cell phones!)
4. An inherent and primary factor affecting detectable vibration velocity level is frequency.
5. Other factors include frame rate, lighting, and surface or target contrast

6. Triggered time-averaging has been found to offer significant improvement in accuracy.

If procedures are used to achieve sufficiently accurate displacement detection, VVA can reliably determine behavior for vibration levels relevant to international standards such as ISO, and as such presents an important addition to a vibration diagnostic tool kit.

References

[1] Timoner, SJ, and Freeman, DM (2001), "Multi-image gradient-based algorithms for motion estimation", Optical Engineering 40.9, pp. 2003-2016

[2] Liu et al (2005), "Motion magnification", ACM Transactions of Graphics (TOG) – Proc. of ACM SIGGRAPH 2005, TOG Homepage, v. 24, Issue 3, July

[3] Wu, H-Y et al (2012), "Eulerian video magnification for revealing subtle changes in the world", ACM Trans. of Graphics (Proc. SIGGRAPH 2012), v. 31, August

[4] Marscher, W.D., and Onari, M.M. (2018), Video Motion Magnification vs. Operating Deflection Shapes for Machinery Diagnosis, MFPT2018, Virginia Beach, May

[5] Chen, J.G.; Davis, A,; Wadhwa, N; Durand, F; Freemans, W.T.; and Oral Büyüköztürk, O., Video Camera–Based Vibration Measurement for Civil Infrastructure Applications, ASCE JIS, 2016

[6] Diamond, D., Oberholster, A., Heyns, S., Structural Vibration Field Test Using Optical Flow Analysis of a Video Sequence, Proc. AIP Conf. , 2016

[7] ISO 10816-3-2018, Mechanical vibration - Evaluation of machine vibration by measurements on non-rotating parts - Part 3: Industrial machines with nominal power above 15 kW …, International Standards Organization, Geneva Switzerland, c. 2009

Where Theory meets Practice

EXPERIMENTAL STUDY ON INTERFACE DEBONDING IMAGINATION FOR A CONCRETE FILLED STEEL TUBE SPECIMEN OF A LONG-SPAN SUSPENSION BRIDGE TOWER WITH EMI

Qian Liu, PhD candidate
College of Civil Engineering, Huaqiao University
Jimei Avenue 668, Xiamen 361021, China

Main research fields: Study on nondestructive testing method and mechanism of concrete-filled steel tube by using electromechanical impedance technique.
Email: hqulq@stu.hqu.edu.cn

Jiang Wang, PhD candidate
College of Civil Engineering, Huaqiao University
Jimei Avenue 668, Xiamen 361021, China

Main research fields: Study on nondestructive testing method and mechanism of concrete-filled steel tube by using guided wave technique.
Email: jiangwang@hqu.edu.cn

Bin Xu, PhD, Professor and Director
College of Civil Engineering, Huaqiao University
Key Laboratory for Intelligent Infrastructure and Monitoring of Fujian Province (Huaqiao University)
Jimei Avenue 668, Xiamen 361021, China

Main research fields: Study on new technology and mechanism of nondestructive testing of civil engineering structures; Identification of structural nonlinear behavior; New wind support structure system; Multi-disaster protection; Multi-scale and multi-physics field simulation of concrete materials and components
Email: binxu@hqu.edu.cn

ABSTRACT:
Concrete-filled steel tube (CFST) structures have been widely used in high-rise buildings and long-span bridges. Interface debonding defects between concrete and steel tube of CFST members decrease their load-carrying capacity, stiffness and ductility. In this study, to visualize the inner and inaccessible interface debonding defects in large-scale CFST members in a steel box-CFST composite tower of a long-span suspension bridge, experimental study on the applicability of an interface debonding detection and visualization approach for a full-scale circular CFST specimen with a diameter of 3.6m using electromechanical impedance (EMI) measurement is carried out. Interface debonding defects with various dimensions are mimicked at different locations at the inner surface of the steel tube of the CFST specimen. Piezoelectric Lead Zirconate Titanate (PZT) patches used as sensors are mounted on the outer surface of the steel tube of the specimen. Based on the shift of the peak frequency of the EMI curve and the corresponding Root Mean Square Deviation (RMSD) result of each PZT sensor measurement, the location of each mimicked interface debonding defect is detected successfully. Moreover, an interface debonding defects imagination method is proposed to visualize the debonding defects. The feasibility of the proposed debonding detection and imagination approach is verified.

KEY WORDS: Concrete-filled steel tube (CFST); Electromechanical impedance (EMI) method; Experimental studies; Interface debonding defects; Non-destructive testing; Steel box-CFST composite tower structure.

INTRODUCTION

Taking advantages of mechanical performance of both steel and concrete, steel-concrete composite structures including concrete filled steel tube (CFST) members have been widely used in high-rise buildings in the past two decades. Recently, a long-span suspension bridge with a main span of 2,300 m is under construction in China. The bridge has a box-CFST composite tower with a height of 350 m. Each circular CFST column employed in the bridge tower has a diameter of 3.6 m. During the construction of the steel box-CFST composite tower, interface debonding defects may occur between steel tube and concrete due to inadequate concrete vibration, unavoidable shrinkage and creep of mass concrete, uneven hydration heat distribution during curing. The interface debonding defects have a negative impact on the overall mechanical performance of the steel box-CFST composite tower structure [1]. It is crucial to develop effective interface debonding detection approach for CFST members.

In recent years, ultrasonic guided waves techniques have been employed to detect concrete cracks, interlayer debonding of composite laminates as typical non-destructive testing (NDT) approaches [2-4]. Due to the electromagnetic shielding effect of the outer steel tube as a metal medium, traditional electromagnetic debonding detection method does not work for internal interface debonding detection of CFST members. Using Piezoelectric Lead Zirconate Titanate (PZT) materials, NDT methods based on Electromechanical impedance (EMI) measurements have been developed for defect detection of engineering structures [5, 6]. Tawie and Lee [7] used the EMI method to detect the hardening strength of concrete during the curing process, and verified the influence of concrete mix ratio on the resonance frequency of EMI measurements. Zhu et al. [8] explored an EMI-based skin-core interface debonding detection technology for honeycomb sandwich composite structure and the degree of debonding is quantified by the defined damage index of root mean square deviation (RMSD) of the EMI measurement. Shao et al. [9] evaluated the bolt pre-tightening force of a structure based on the EMI frequency shift. Xu et al. [10] used the EMI method to detect the interface debonding defects of a large irregular multi-chambers steel reinforced concrete (SRC) column and the feasibility of this method was verified experimentally. Wen et al. [11] used the multi-physical field coupling finite element model to study the feasibility of cavity detection approach for CFST members based on EMI.

Further experimental studies on the feasibility of EMI based interface debonding detection and visualization method for large-scale CFST members are desired in practice. In this paper, an experimental study on the feasibility of EMI measurement-based interface debonding defects detection and visualization approach for a large-scale CFST member of a long-span suspension bridge tower was conducted. Interface debonding defects at different locations at the inner surface of the steel tube of the specimen are mimicked and number of PZT patches are mounted on the outer surface of steel tube as sensors. The EMI of each PZT sensor at defined frequency regions is measured and compared. The influence of interface debonding defects of different sizes on the EMI measurements of PZT sensors was analyzed and the location of each mimicked interface debonding defect was determined. Moreover, the distribution of the interface debonding defects of the CFST specimen was visualized using an interpolation algorithm based on the RMSD results.

1 INTERFACE DEBONDING DETECTION PRINCIPLE BASED ON EMI MEASUREMENT WITH SURFACE MOUNTED PZT SENSORS

The most significant advantage of the EMI method for defect detection is that both the positive and negative piezoelectric effects of PZT patches are used simultaneously. A PZT patch can be installed on the outer surface of a structure as both actuator and sensor to identify defects. A one-dimensional coupled system proposed by Liang et al. as shown in Fig. 2 can be employed to describe the principle of the EMI based defect detection for structures [12]. The PZT patch is simplified to a single-degree-of-freedom system considering only the axial compression deformation and interacting with the host structure along the length of the PZT patch. The electrical impedance of the surface-mounted PZT sensor is coupled with the mechanical impedance of the host structure. For CFST members, when interface debonding defects occur between steel tubes and core concrete, the local stiffness at the location of the debonding defect has a decrease, which results in a change in the local mechanical impedance of the CFST and then a difference in the EMI signature measured directly by the coupled PZT patch. Therefore, by comparing EMI signature before and after the interface debonding damage occurs in the host structure, the interface defect can be identified.

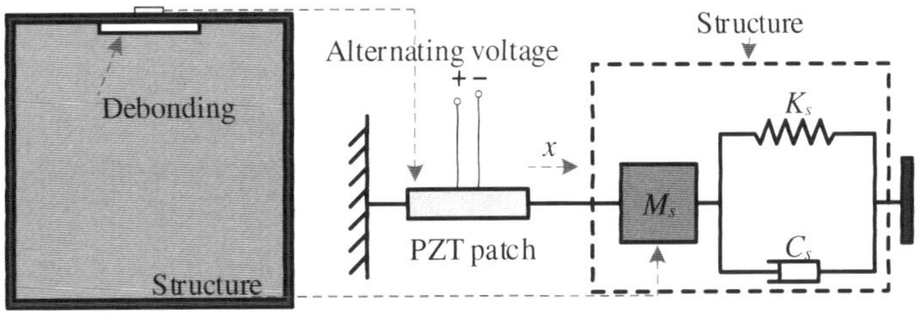

Fig. 2 EMI measurement system with one-dimensional PZT-structure coupling model

2 EXPERIMENTAL STUDY ON INTERFACE DEBONDING DEFECTS DETECTION OF A FULL-SCALE CFST MEMBER

2.1 FULL-SCALE CFST SPECIMEN

The full scale CFST specimen is identical to the original design of the CFST member in the steel box-CFST composite tower of the suspension bridge with a main span of 2,300 m. The CFST members are filled with C60 self-compacting concrete, and the steel tube is equipped with horizontal diaphragm, two vertical partition plates mutually perpendicular to each other and shear studs to ensure the composite effect between the steel tubes and the concrete. The full-scale circular CFST member has a diameter of 3.6 m and a height of 0.6 m. The thickness of the steel tube, the diaphragm, the vertical plates are 20 mm. The steel tube is divided equally into four areas A, B, C, and D by the vertical partition plates as shown in Fig. 3.

(a) CFST specimen (b) Cross section diagram

Fig. 3 Full-scale CFST member on site before concrete pouring

2.2 ARRANGEMENT OF ARTIFICIALLY MIMICKED INTERFACE DEBONDING DEFECTS

To mimic interface debonding defects that might occur between the inner surface of the steel tube and concrete core of the CFST specimen, number of Ethylene-Vinyl Acetate (EVA) foam plates are pasted on the inner surface of steel tube at designed locations with epoxy resin before concrete pouring. Fig. 4 shows the EVA foam material and the mimicked interface debonding defects installed on the inner surface of the steel tube of the CFST specimen.

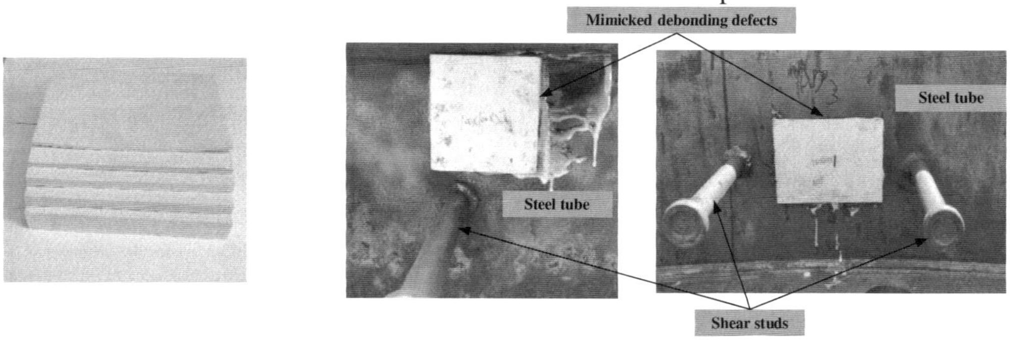

(a) EVA foam material (b) Mimicked interface debonding defects

Fig .4 Mimicked interface debonding defects in CFST specimen

In this experimental study, the mimicked interface debonding defects are arranged at two different cross sections, Section A and Section B, of the CFST specimen. Section B is under the horizontal diaphragm where debonding occurs easily. The dimensions of each mimicked debonding defect are shown in Table 1 in details. The locations of all mimicked debonding defects are shown in Fig. 5. After insulation and waterproof treatment, six PZT patches are pasted on the locations of the center of the mimicked debonding defects on the outer surface of the steel tube as sensors. For comparison, six PZT sensors labelled as AH1-3 and BH1-3 are placed at healthy locations on the two sections where no interface debonding defect is mimicked and are represented by cross lines in Fig. 5. To detect the region of each interface debonding in details, five PZT sensors are installed on the outer surface of the steel tube corresponding to each defect. Taking the interface debonding defect AD1 as an example, the PZT sensor at the center of the defect is labelled as AD1-1, and other four PZT sensors, named as AD1-2, AD1-3, AD1-4 and AD1-5, are located at 2 cm outside of the four defect edges, as shown in Fig. 6.

Table 1 Design of interface debonding defects

Label Number	Size (Radial ×Circumferential × Vertical, mm ×mm × mm)	Defect location
AD1	1×50×100	Section A
AD2	1×100×100	Section A
AD3	1×150×100	Section A
BD1	3×100×100	Section B
BD2	5×100×100	Section B
BD3	7×100×100	Section B

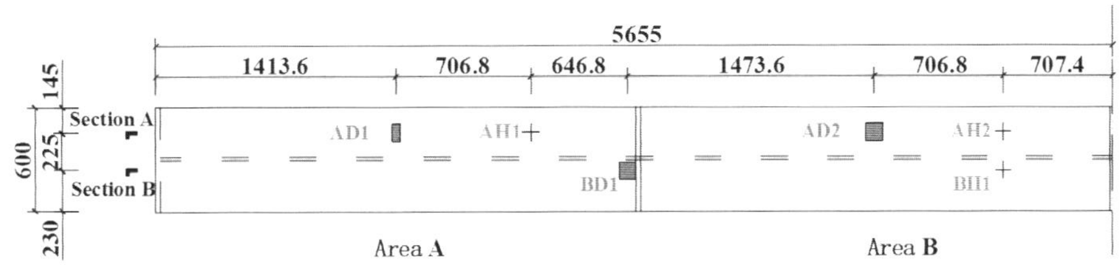

(a) Areas A and B

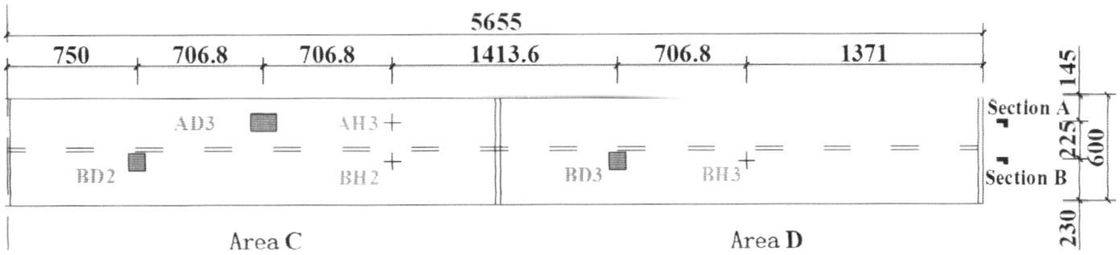

(b) Areas C and D

Fig. 5 Locations of mimicked defects and the PZT sensors at healthy regions at the expended view of the steel tube (unit: mm)

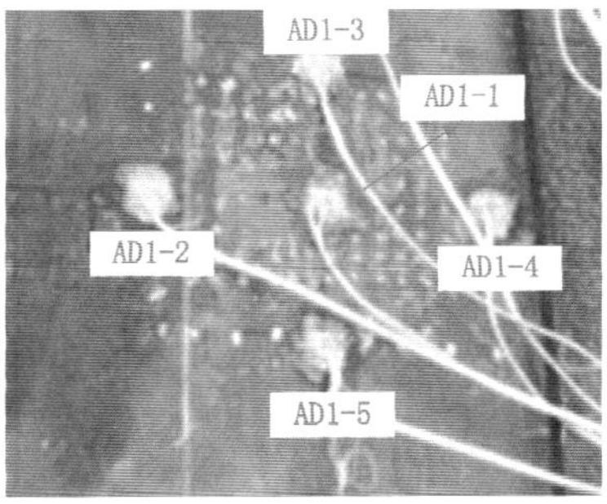

Fig. 6 PZT sensors arrangement at defect AD1

Fig. 7 EMI measurement test setup

3 TEST RESULT ANALYSIS

3.1 IMPEDANCE SIGNATURE CURVE ANALYSIS

The EMI of each PZT sensor at a selected frequency band is measured with an impedance analyzer as shown in Fig. 7. A suitable frequency band is selected [13].The comparison of the EMI measurements of all PZT sensors at the center of each mimicked debonding defect at Section A and the average of the EMI measurements of all the three PZT sensors at healthy locations are made and shown in Fig. 8. It can be seen from Fig. 8 that the impedance curve of average of the three PZT sensors (AH1, AH2, AH3) at healthy locations is smooth. However, the impedance curves of the three PZT sensors (AD1-1, AD2-1, AD3-1) located at the center of three mimicked debonding defects have obvious fluctuations and new peak occurs. Moreover, with the increase of the debonding defect dimension, left-shift of the peak frequency can be found. It suggests that the interface debonding defect can be detected successfully based on the EMI measurements on the selected frequency band. Fig. 9 shows the results in section B and similar findings can be found.

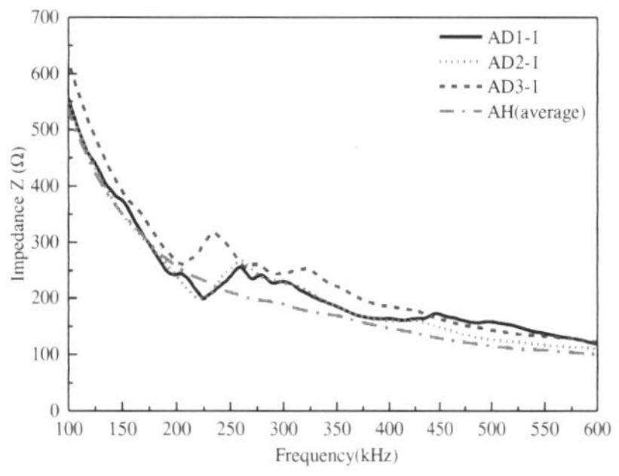

Fig. 8 Impedance of PZT sensors in section A

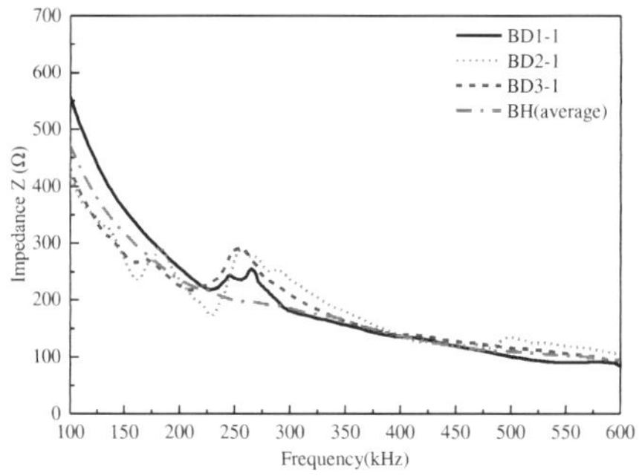

Fig. 9 Impedance of PZT sensors in section B

3.2 INTERFACE DEBONDING VISUALIZATION USING DAMAGE INDEX

The RMSD of the EMI measurements of the *j*-the PZT sensor, defined in Equation (1), is used to visualize the location of the mimicked interface debonding defects of the CFST specimen [13].

$$RMSD_i = \sqrt{\frac{\sum_{j=1}^{N}(Z_{i,j}^1 - Z_{i,j}^0)^2}{\sum_{j=1}^{N}(Z_{i,j}^0)^2}} \times 100\% \quad (1)$$

where $Z_{i,j}^1$ represents the impedance test data of the *i*-th PZT sensor located at a region to be detected at the *j*-th frequency, N is the total number of frequency points, and $Z_{i,j}^0$ is the average of the impedance of PZT sensors at healthy locations at the *j*-th frequency.

The EMI measurements collected from all the PZT patches at the two sections of the specimen were used for RMSD calculations, and the results are shown in Fig. 10. It can be seen from Fig. 10 that the RMSD values of the PZT sensors installed at the location of debonding defects are significantly higher than them at the healthy locations. The RMSD values of the measured points at the location without defect are less than 5%. The RMSD values of the five PZT sensors around the center of the mimicked defect AD1 are shown in Fig. 11. It can be seen that the RMSD value corresponding to PZT sensor AD1-1 installed at the center of the mimicked defect is significantly higher than them of all other four PZT sensors at locations of 2 cm outside of the defect edge. RMSD values of the regions covered by the five PZT sensors around each mimicked interface debonding defect are determined by an interpolation method based on the RMSD values of the five PZT sensors. The region out of the tested locations are assumed to be healthy. The RMSD value matrix at different location is mapped to a color matrix using the colormap function. The location and scope of the mimicked interface debonding defects and the comparison with the actual location shown in solid line in the expended view of the steel tube can be visualized as shown in Figs. 12-13. The mimicked interface debonding can be detected and visualized successfully.

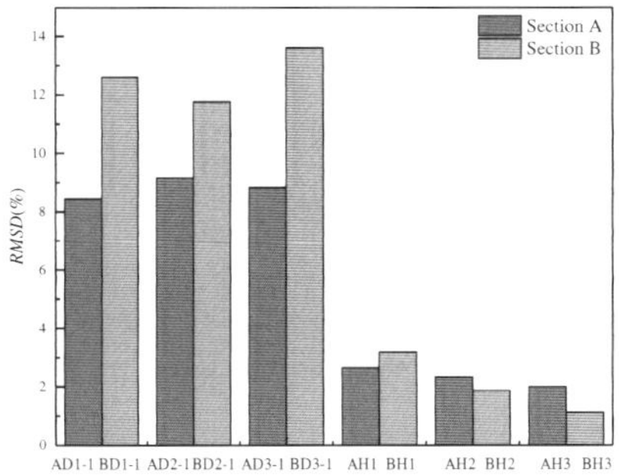

Fig. 10 RMSD values for each section

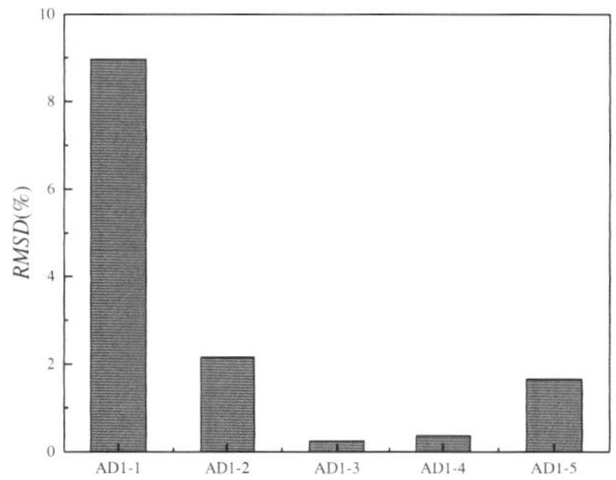

Fig. 11 RMSD values of five sensors around debonding defect AD1

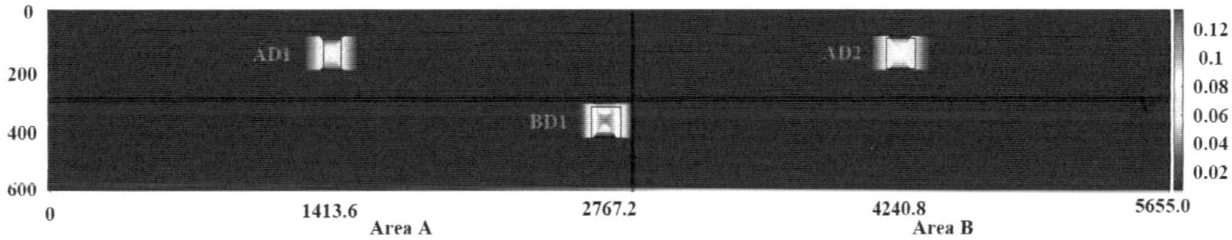

Fig. 12 Imaging of interface debonding defects (Areas A and B)

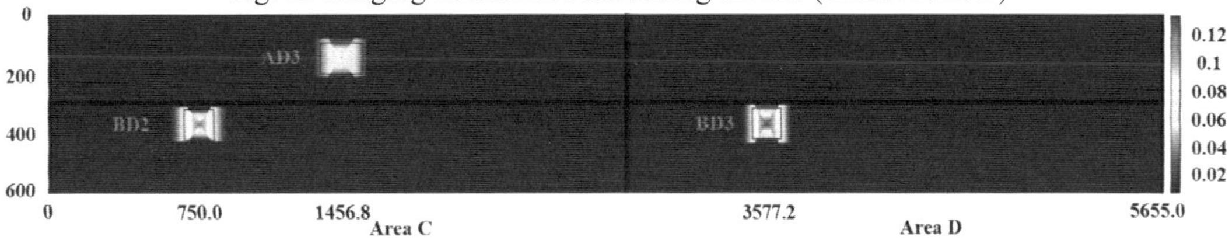

Fig. 13 Imaging of interface debonding defects (Areas C and D)

CONCLUSION:

In this paper, the feasibility of an interface debonding defect detection and visualization approach for a large-scale CFST specimen using the EMI measurement is experimentally verified and the following conclusions can be made.

(1) The existence of interface debonding defects makes the impedance curve fluctuate significantly and the frequency of the peak of the EMI has a left-shift. The EMI measurements of PZT patches at healthy locations are consistent with each other.

(2) By the comparison the EMI measurements of surface-mounted PZT patches, the existence of interface debonding defects at two different cross sections can be detected successfully. Using an interpolation algorithm based on the EMI measurements around a mimicked interface debonding defect, the distribution of the interface debonding defects can be visualized and the results are close to the artificially mimicked interface debonding defects.

REFERENCES:

[1] Nematzadeh, M. , Fazli, S., Naghipour M., et al, "Experimental study on modulus of elasticity of steel tube-confined concrete stub columns with active and passive confinement", Engineering Structures, 2017, (130): 142-153.
[2] Lee, H., Koo, B., Chattopadhyay, A., et al, "Damage detection technique using ultrasonic guided waves and outlier detection: Application to interface delamination diagnosis of integrated circuit package", Mechanical Systems and Signal Processing, 2021, 160: 107884.
[3] Pavlopoulou, S., Worden, K., Soutis, C., "Novelty detection and dimension reduction via guided ultrasonic waves: Damage monitoring of scarf repairs in composite laminates", Journal of Intelligent Material Systems and Structures, 2016, 27(4): 549-566.
[4] Rautela, M., Gopalakrishnan, S., "Ultrasonic guided wave based structural damage detection and localization using model assisted convolutional and recurrent neural networks", Expert Systems with Applications, 2021, 167: 114189.
[5] Antunes, R.A., Cortez, N.E., Gianesini, B.M., et al, "Modeling, simulation, experimentation, and compensation of temperature effect in impedance-based SHM systems applied to steel pipes", Sensors, 2019, 19(12): 1-23.
[6] Su, Y.F., Han, G., Nantung, T., et al, "Novel methodology on direct extraction of the strength information from cementitious materials using piezo-sensor based electromechanical impedance (EMI) method", Construction and Building Materials, 2020, 259: 119848.
[7] Sun, F.P., Chaudhry, Z., Liang, C., et al, "Truss structure integrity identification using PZT sensor-actuator", Journal of Intelligent material systems and structures, 1995, 6(1): 134-139.
[8] Zhu, J., Wang, Y., Qing, X., "Modified electromechanical impedance-based disbond monitoring for honeycomb sandwich composite structure", Composite Structures, 2019, 217: 175-185.
[9] Shao, J., Wang, T., Yin, H., et al, "Bolt looseness detection based on piezoelectric impedance frequency shift", Applied Sciences, 2016, 6(10): 298.
[10] Xu, B., Li, J., Li, Y.H., et al, "Detection of Debonding Defect of Irregular SRC Column Interface Based on PZT Impedance Measurement", Nondestructive Testing, 2016, 38(09): 29-32.
[11] Wen, J.Y., Xu, B., Wang, H.D., "Study on Numerical Simulation of Piezoelectric Impedance Detection of Composite Concrete Structure Defects", Piezoelectrics and Acoustooptics, 2018, 40(02): 276-279.
[12] Liang, C., Sun, F.P., Rogers, C.A., "An impedance method for dynamic analysis of active material systems", Journal of Intelligent Material Systems and Structures, 1994,116(1): 120-128.
[13] Kuang, J., Xu, B., "Local damage detection for steel rebar by impedance measurements of PZT sensors", Third International Conference on Smart Materials and Nanotechnology in Engineering, Shenzhen, China, 2011.

ELECTRICAL SIGNATURE ANALYSIS OF OIL AND DRY TYPE TRANSFORMERS

Howard W Penrose, Ph.D., CMRP
President, MotorDoc LLC
howard@motordoc.com

Howard Penrose is the president of MotorDoc LLC, 2021-2024 chair wind power standards American Clean Power, a past chair of the Society for Maintenance and Reliability Professionals, and a past chair IEEE Chicago Section. He serves on the IEEE DEIS and PES Electrical Machinery Subcommittees. Dr. Penrose is the leading researcher in Electrical Signature Analysis and related applications from commercial/industrial to utility and mobile equipment to military applications. He is also a leading researcher and designer of electrical machinery for special applications. Dr. Penrose also leads research into specialty machine learning, data science and AI applications for commercial/industrial/utility and energy efficiency.

ABSTRACT: Periodic and continuous electrical signature analysis can be used to evaluate the condition of both oil and dry-type transformers. In this paper we will discuss how such conditions as internal and external loose connections, resonance, looseness, and failing electrical components can be detected. Case studies from several applications including wind power, solar, and industrial dry and oil-type transformers will be presented.

KEY WORDS: transformer, electrical signature analysis, ESA, MCSA, power quality, electrical reliability

INTRODUCTION:
The purpose of Electrical Signature Analysis (ESA) is to use the magnetic field in the air gap of an electrical machine to evaluate power quality, the condition of the electrical and mechanical components in an electric motor or generator and driven equipment. The analysis is performed using measured voltage and current data with the line frequency acting as the amplitude modulated carrier frequency. Sample rates, FMAX, bin size, and Nyquist are similar to that of vibration analysis with several exceptions including the Nyquist value is 2, and the data being analyzed in an FFT spectra is either in linear data or decibels. Decibel levels are read 'down' from the peak voltage or current to associated peaks and are presented as -dB. Unlike in vibration where a value such as stator conditions are the number of stator slots times the running speed, ESA is the number of stator slots times the RPM +/- the line frequency (and sometimes mulitples).

In the case of electric machines most conditions and the forcing functions associated with them are interpreted directly from the supplied data. This means that a defect in a bearing, rotor or stator, are explicit while the use and measurements of the transformer are implicit and infer conditions associated with transformer condition and reliability. With transformers we review phase impedance, power factor, phase balance, harmonics, and variations in the magnetic fields between primary and secondary.

Figure 1: Oil Filled Power Transformer

Figure 1 represents an oil filled power transformer. The posts on the top are called 'bushings,' which provide an insulated distance for power cabling and seal the oil in. The fins on the sides are radiators for cooling the transformer oil. The tank on top is an expansion tank for oil that expands as it heats up. There will also be a blanket of nitrogen on top of the oil to prevent water contamination, which would damage insulating materials.

Figure 2: Dry-type transformer in an oven for drying.

The primary difference between an oil-type transformer and dry-type transformer is the cooling medium and resulting size. Dry-type transformers, such as in Figure 2, use air as the cooling medium. Both require slightly different maintenance tasks and have slightly different failure modes. In the case of oil-type transformers the cooling medium can also accelerate reliability issues through degradation from soluble gasses generated from age, contamination, and operation. Dry-type transformers are subject to contamination and problems with air flow.

Both transformer types are subject to problems associated with power quality, loose connections, cooling medium issues, and other conditions associated with loads and environment. There are a number of existing technologies used for different voltage levels and fault types. These are normally passive or injected technologies such as turn to turn ratio testing, oil analysis, ultrasound, vibration analysis, partial discharge testing, insulation resistance, and others. There have been a number of papers and articles associated with the use of electrical signature analysis as well as any number of references associated with power quality testing. For the purpose of this paper we will focus on the types of issues associated with ESA fault detection in transformers.

The use of ESA for transformers adds an additional level for fault detection, especially when used for continuous monitoring. Conditions we will cover within this paper do not include all applications, but will cover:

- Power quality, including power factor and harmonics;
- Core excitation in wind power applications;
- Load balancing in solar applications;
- 13.8kV to 480V transformer overload due to ground/neutral harmonic content;
- Loose connection on transformer bushing.

POWER QUALITY:

Figure 3: 125kV to 13.8kV transformer with low power factor and high harmonic loads.

During an electrical reliability evaluation of a food processing site, it was noted that the main transformer supplying the facility (Figure 3) had a 'tinny' ring to it. An evaluation of the subs fed from the transformer, observation of the transformer heating on a 34F day while operating at a measured 70% load, and a plant power factor under 0.8 indicated that the transformer was overloaded and the harmonic and power factor were outside of the operating capabilities of the transformer. Figure 4 is an example of the primary side of one of the sub transformers and Figure 5 is the current on ground leading back from the sub. Each of the subs add to the power factor, harmonic and ground/neutral loads at the utility feed.

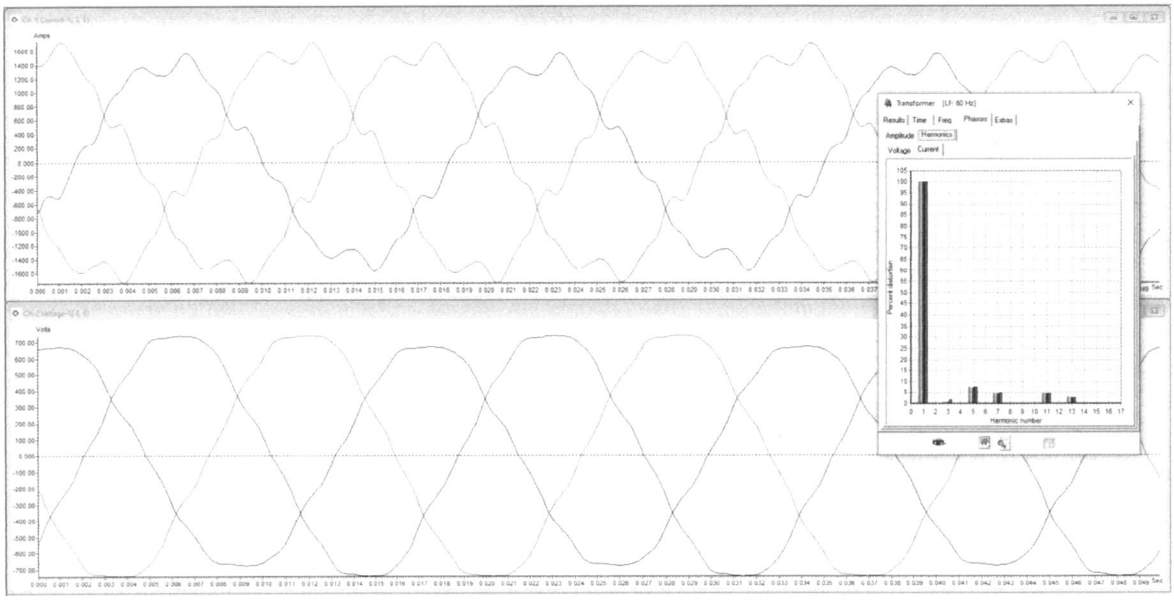

Figure 4: Secondary of sub-transformer with primarily 5th and 7th harmonic and mulitples in voltage and current.

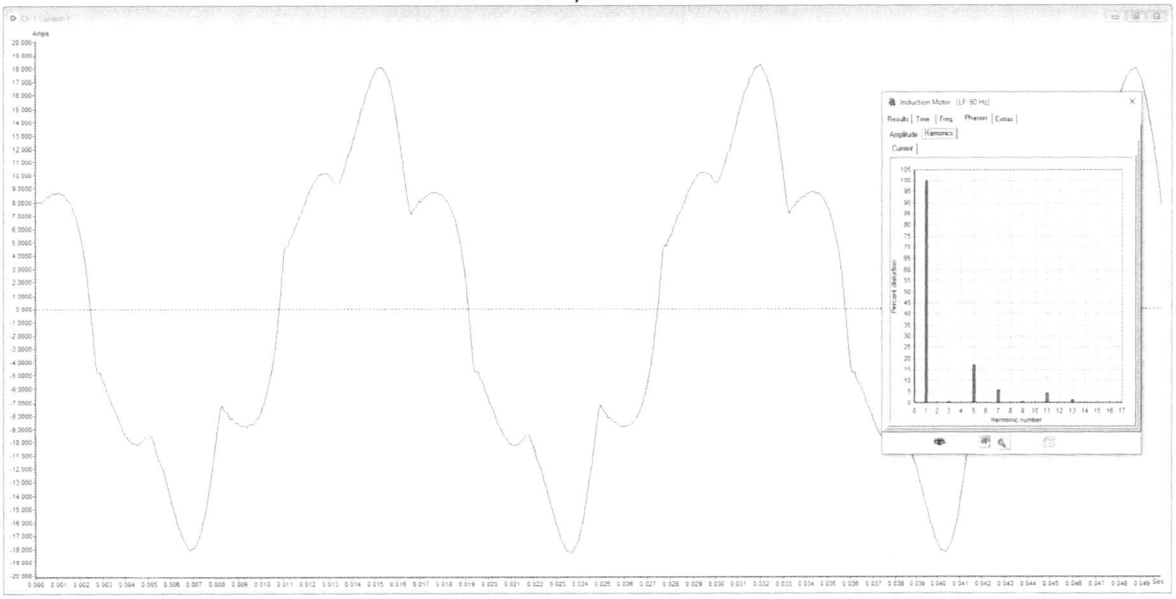

Figure 5: Ground current at sub transformer (note: this value should be in milli-Amps)

The harmonic content is found in both power and ground being fed back to the transformer as well as poor power factor add to core heating and the overall transformer effective load. The 70% load quoted from the utility is based upon the apparent load and does not include derating and core resonance due to poor power factor and harmonic content.

Due to access restrictions a smart phone audio was obtained and processed to show power loss (Figure 6) and decibel patterns (Figure 7). The purple squares in Figure 6 are related to loose components and the red arrow points to core vibration due to harmonics and high ground currents. Figure 7 shows power factor and core vibration (ref. IEEE Std C57.136-2000). The power quality conditions will reduce the life of the transformer.

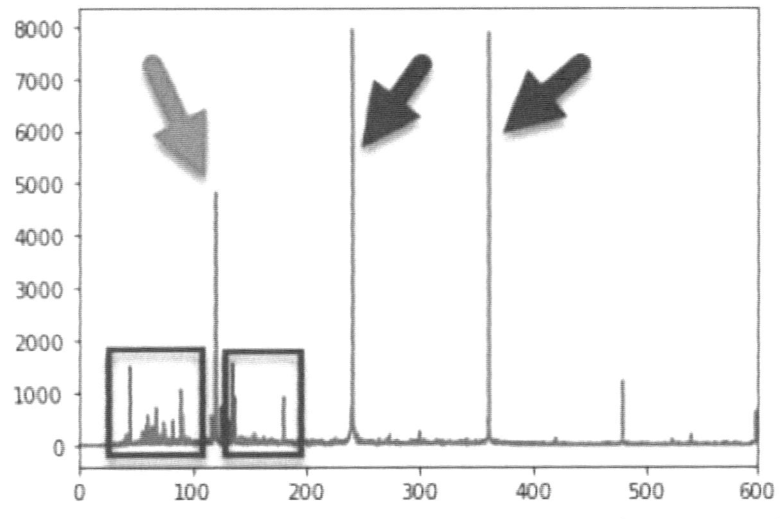

Figure 6: power analysis of audio related to transformer sound.

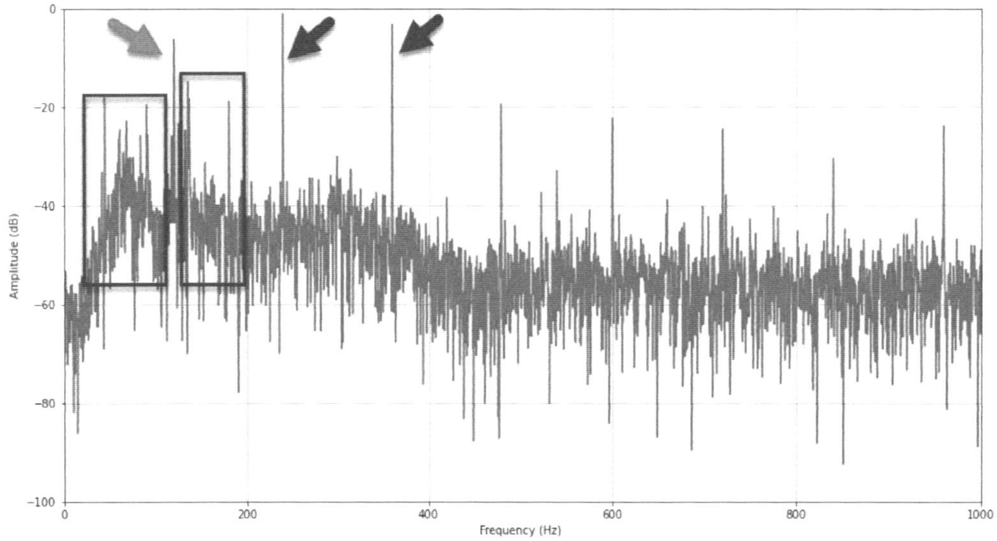

Figure 7: decibel analysis of transformer audio. 120 Hz and related harmonic are directly related to power factor and harmonic content.

CORE EXCITATION IN WINDPOWER:

A common problem in the wind industry is overheating transformers which is due to a combination of factors including the firing frequency of rotor inverters in Doubly-Fed Induction Generators (DFIG). There are a number of factors that generate this issue that are outside of the scope of this paper. However, the conditions are easily detectable with ESA to include the detection of the inverter frequency and a transformer excitation frequency that falls between the first and second inverter harmonic.

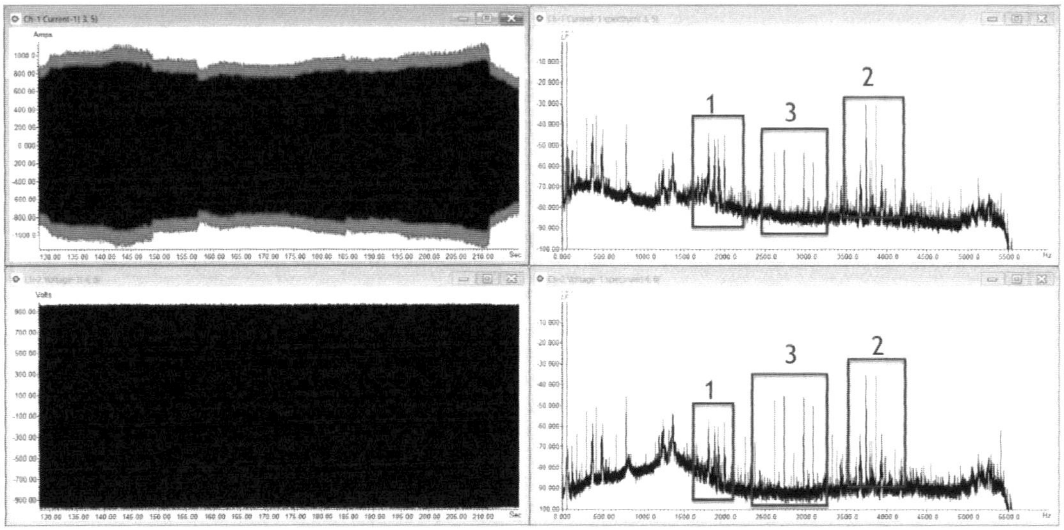

Figure 8: ESA voltage and current data on a wind generator. 1 is the rotor inverter frequency, 2 is the second harmonic of the rotor inverter frequency, and 3 is related to transformer core resonance.

The frequencies present in Figure 8 generate heating in the core of the transformer and will generally increase thermal gassing, depending on the dielectric oil being used, which primarily consists of hydrogen. This will generally thermally and chemically degrade insulating materials in the transformer. In dry-type transformers resonance will also effect connections.

LOAD BALANCING IN SOLAR APPLICATIONS:

One of the primary concerns in solar applications is balancing the load between the two lower voltage primaries and the secondary. Figure 9 is a solar transformer with an unbalanced primary and inverter harmonics which convert the solar DC power to AC at the transformer. The primaries are often phase shifted to reduce output harmonics in the distribution system as shown in the simple transformer sketch in Figure 10.

Figure 9: Solar transformer being tested and evaluated between both primaries from the output of the solar inverter.

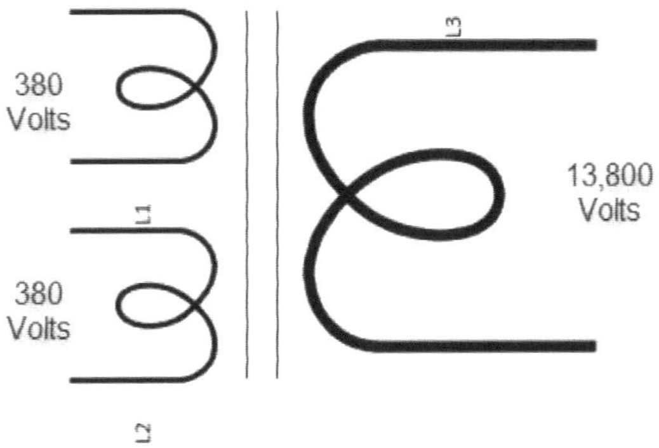

Figure 10: Rough circuit sketch of a solar step-up transformer.

When unbalances exist, the firing frequencies and transformer core resonance will show in similar fashion as with wind turbines as shown in Figures 11 and 12.

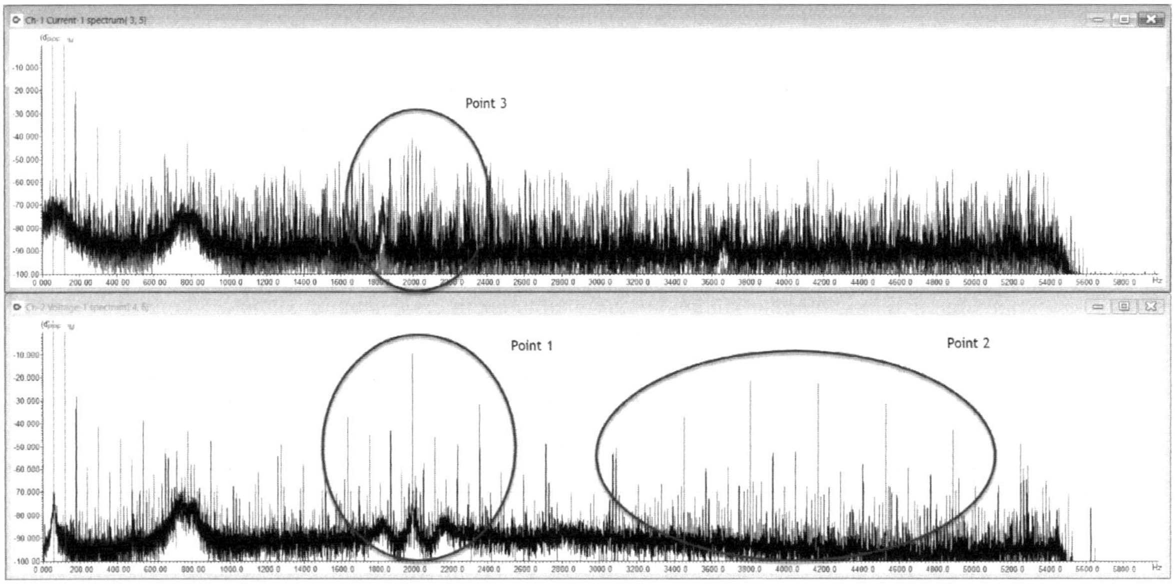

Figure 11: Spectral analysis of one primary of solar transformer. Points 1 and 3 represent the inverter frequency. Point 2 represents the resonance present in the transformer.

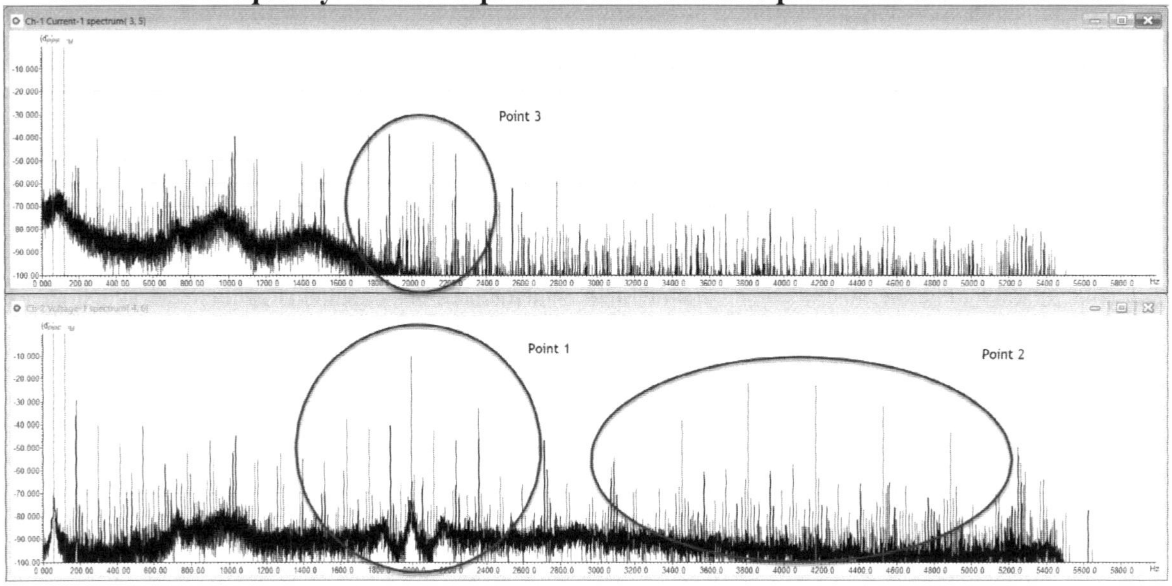

Figure 12: Spectral analysis of the other primary in the same transformer.

Figures 13 and 14 are of the electrical and power data associated with the solar transformer. As is observed, the transformers are heavily unbalanced between the sides. In cases where the transformer primaries are balanced, the resonance rarely appears to occur. This condition results in heating and gassing with reduced insulation and oil life.

Transformer [LF: 60 Hz]

Results | Time | Freq. | Phasors | Extras

	Phs-1	Phs-2	Phs-3	Total	Units
Power factor	0.997	-0.427	-0.438	0.630	
Real Pwr.	27.1	9.1	12.9	49.1	KW
Reactive Pwr.	2.0	19.2	26.5	47.7	KVAF
Apparent Pwr.	27.2	21.2	29.4	77.9	KVA
Running Cnt.	111.2	87.0	121.3	106.5	Amp
Line Voltage	397.8	395.8	397.3	396.9	Volt

Horse Power

Figure 13: The electrical data associated with one of the primaries. Note the 49.1 kW and current values.

Transformer [LF: 60 Hz]

Results | Time | Freq. | Phasors | Extras

	Phs-1	Phs-2	Phs-3	Total	Units
Power factor	1.000	-0.494	-0.347	0.631	
Real Pwr.	208.6	91.6	62.8	363.1	KW
Reactive Pwr.	1.9	161.2	170.0	333.1	KVAF
Apparent Pwr.	208.6	185.4	181.2	575.3	KVA
Running Cnt.	869.1	771.2	759.6	800.0	Amp
Line Voltage	393.8	392.7	392.7	393.1	Volt

Horse Power

Figure 14: Electrical data associated with the other primary. Note the 363.1 kW and operating current.

13.8KV TO 480 VOLT TRANSFORMER OVERLOAD DUE TO GROUND/NEUTRAL HARMONIC CONTENT:

The ground and neutral feed back to the transformer and in cases where voltage and current is present, in particular where there are harmonics over the fundamental, will generate heating and increase the loading on the transformer.

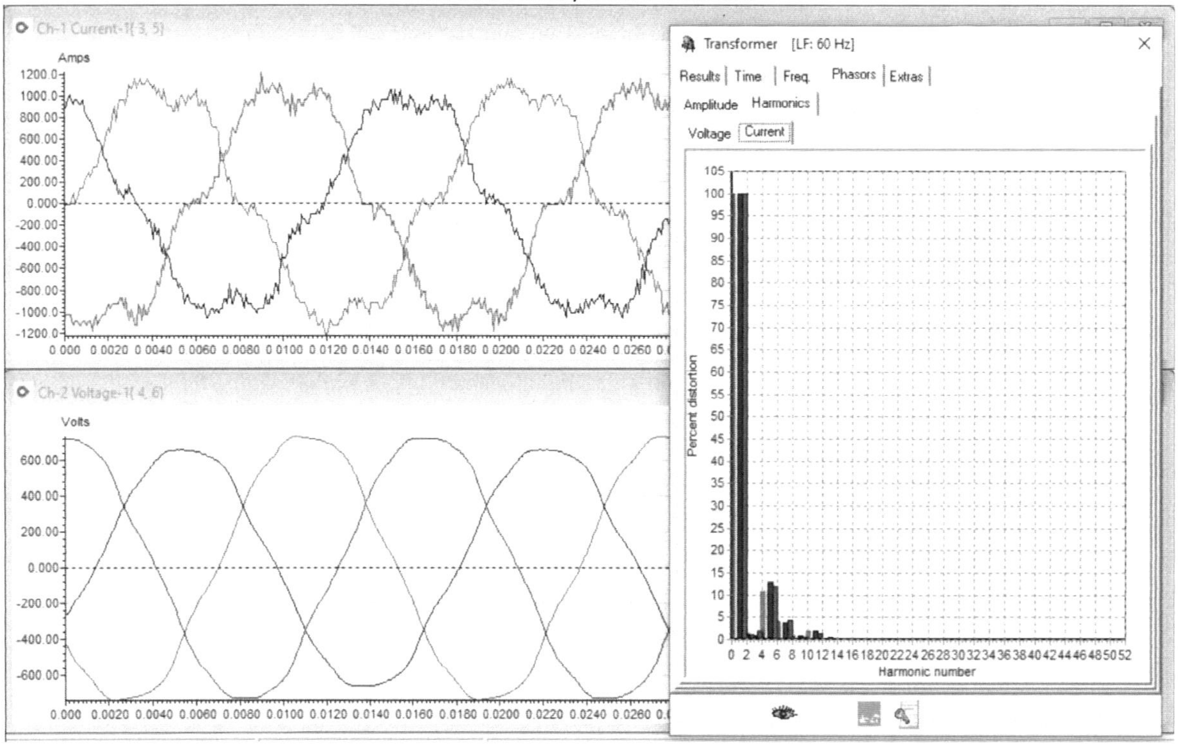

Figure 15: Before neutral and ground harmonic correction note even and odd harmonics.

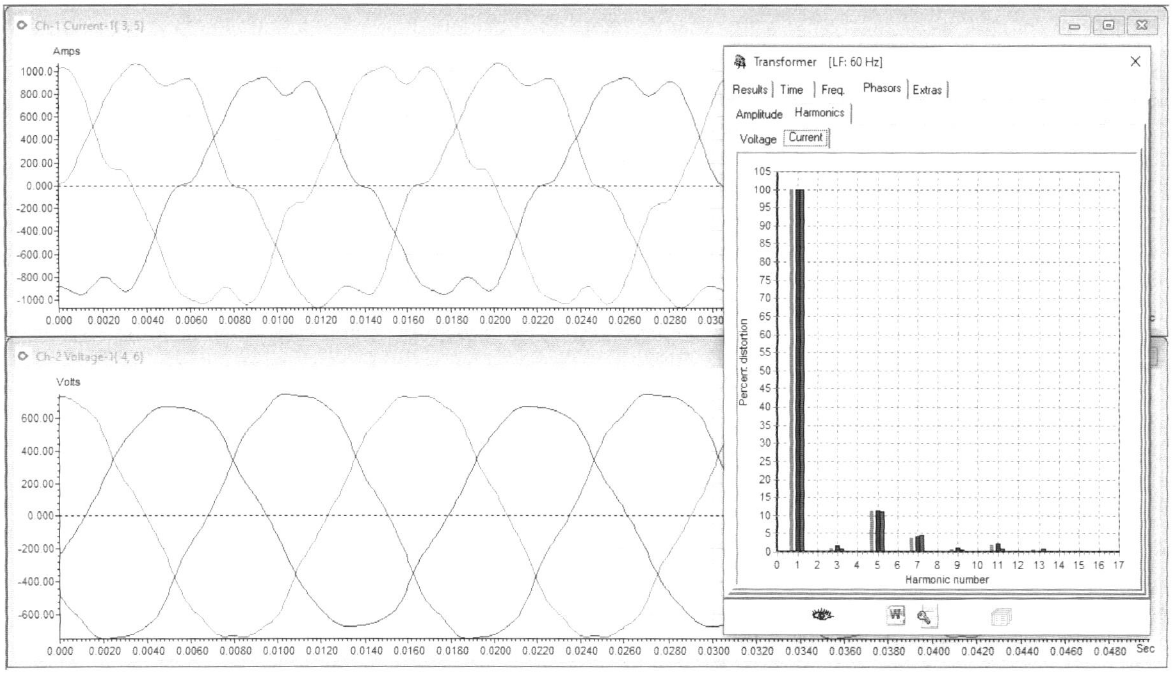

Figure 16: After installation of neutral and ground harmonic correction.

Note the changes between Figures 15 and 16 where neutral and ground harmonic correction was implemented. The noise in the current waveform decreases and the even harmonics are eliminated due to a reduction in load on the transformer core. Note the comparison is at a similar load current.

LOOSE CONNECTION ON TRANSFORMER BUSHING

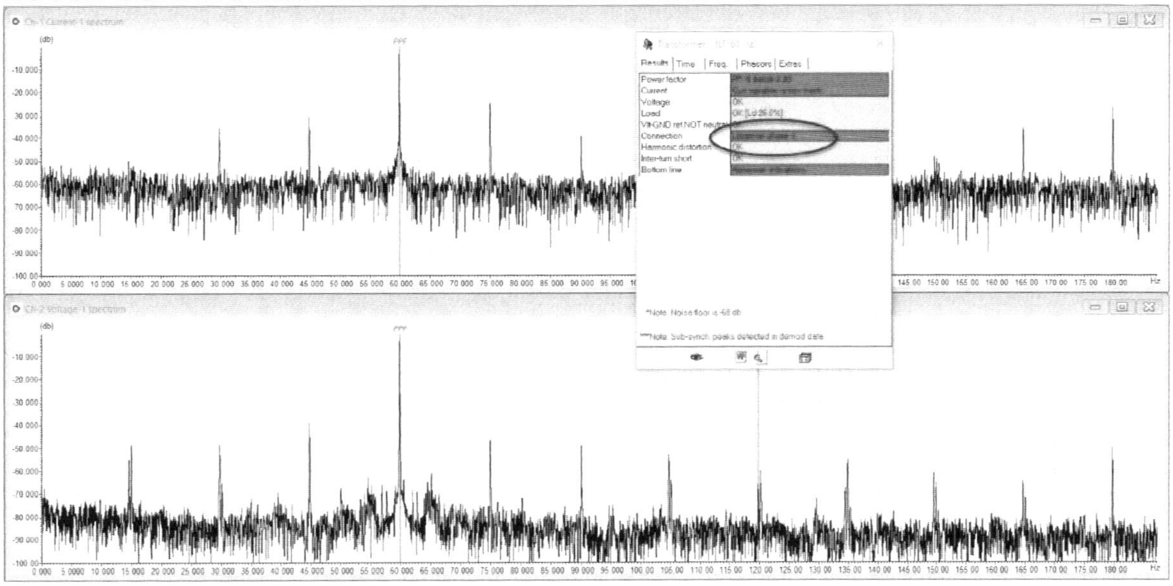

Figure 17: Loose connection found on lighting transformer.

Transformer [LF: 60 Hz]

Results | Time | Freq. | Phasors | Extras

	Phs-1	Phs-2	Phs-3	Total	Units
Power factor	0.787	0.652	0.784	0.739	
Real Pwr.	49.3	39.1	39.8	128.2	KW
Reactive Pwr.	38.7	45.6	31.5	115.8	KVAR
Apparent Pwr.	62.6	60.1	50.8	173.5	KVA
Running Cnt.	9.2	8.8	7.4	8.4	Amp
Line Voltage	11890	11904	11827	11873	Volt

Horse Power

Figure 18: Impedance unbalance found in phase C.

Loose connections are found through a combination of impedance unbalance and will often have even harmonics present. The impedance unbalance distinguishes between harmonic loading and the loose connection.

CONCLUSION:

ESA is an excellent tool for direct or incidental detection of major defects in transformers by evaluating the power quality, phase impedance, and impact of loads. In the examples provided in this paper we have been able to detect: loose connections, overloading, neutral and ground impacts, power factor problems, and a few others. When used as part of continuous monitoring these conditions can be found early enough to mitigate the defects.

BIBLIOGRAPHY:

Penrose, H. W., Practical Electrical and Current Signature Analysis of Electrical Machinery and Systems, Success by Design Publishing, Lombard, IL, 2022.

MFPT ANNUAL CONFERENCE

Where Theory meets Practice

Vibration and Acoustic Resonance Spectrometry (VARS) for Advanced Streaming Prognostics of Drivetrain System

Matthew Gerdes, Guang C. Wang, Ruixian Liu, Kenny Gross
Oracle Corporation
San Diego, CA 92121, USA

Yawen Wang, Xinqi Wei
The University of Texas at Arlington
Arlington, TX 76019, USA

Matthew Gerdes received his BSc degree from the California State University Northridge in 2017, and a M.Sc. degree from University of California San Diego in 2020. He is currently an ML researcher at Oracle Labs and his current research interests include time series, anomaly detection, machine learning, and vibration analysis.

Yawen Wang is an assistant professor of research at University of Texas at Arlington. He is the director of Vibro-Acoustics and Sound Quality Research Laboratory and Hypoid and Bevel Gear Mesh and Dynamic Modeling Consortium. His research interests include gear noise/vibration, vehicle NVH (noise, vibration and harshness), vibro-acoustic modeling, and data-driven design and diagnostics.

Xinqi Wei is a PhD student at the University of Texas at Arlington. He has been researching on the hypoid gear design and manufacturing. His current research interests include hypoid tooth contact analysis, dynamic analysis, and optimization.

MFPT
ANNUAL CONFERENCE
Where Theory meets Practice

Guang Chao Wang received his M.Sc. degree in Aerospace Engineering at The Ohio State University, and earned his Ph.D. in Energy Science from UC San Diego. He is currently a Machine Learning Researcher in Oracle Labs and has extensive experiences in time series anomaly detection. His other research interests include time series forecasting, distributed energy resources planning, and photovoltaic power modeling.

Ruixian Liu received his M.Sc degree from University of California San Diego in 2019. He is currently a Ph.D student in the Electrical and Computer Engineering department at University of California San Diego. He is also a research assistant at Oracle Labs focusing on machine learning based time series anomaly detection algorithms. His other research interests include machine learning based dynamical systems identification and prediction.

Kenny Gross is a Machine Learning Architect in Oracle Labs. Kenny has introduced numerous advanced pattern recognition and prognostic machine learning innovations that are now in use for Internet-of-Things applications in the fields of utilities, avionics, manufacturing, enterprise data centers, and life sciences. Kenny has 294 US patents issued and pending on his innovations, 226 scientific publications. Kenny's Ph.D. is in nuclear engineering from the University of Cincinnati.

ABSTRACT

Cloud computing providers are increasingly offering Machine Learning (ML) solutions so that large-scale IoT customer can process sensor telemetry signals for predictive and prescriptive maintenance applications. Recently, substantial reductions in costs have initiated a growing trend in the adoption of accelerometers for detection of the onset of mechanical and hydraulic failures. The challenge for Cloud services is sampling rates for accelerometers are beyond bandwidth constraints for direct ingestion into cloud tenancies. In this investigation, the Vibration Acoustic Resonance Spectrometry (VARS) technique is proposed to lower compute costs and increase prognostic prediction accuracy. The VARS converts frequency-domain waveforms into a cluster of correlated and lossless down-sampled time series signals. . Subsequently, an ML signal-processing algorithm called the Multivariate State Estimation Technique (MSET) is used to analyze the signals and detect the incipience of degradation in engineering assets. The experimental testing is conducted on a drivetrain diagnostic simulator, which emulates machinery faults that develop in production grade systems. Various “ground truth” fault types are introduced individually or in combination in a laboratory-controlled environment to conduct rigid assessment of fault detection. The analysis results demonstrate that the proposed algorithm can achieve fast detection of early anomaly detection and high accuracy detection of slowly evolving degradation modes even when the vibrational disturbances are below the noise floor.

KEY WORDS: Vibration; prognostics; machine learning; frequency spectrum.

1. INTRODUCTION:

Historically, accelerometers and acoustic sensors have been used to detect problems in assets in many types of industries problems that result in higher vibration levels, which can be precursors to failure [1–6]. Perhaps the greatest challenge for conventional condition-based surveillance of assets is the inherent "seesaw" effect between sensitivity for catching incipient degradation mechanisms and false alarms. This is a consequence of placing thresholds limits on Power Spectral Density (PSD) peak amplitudes [7–10]. Monitoring system designers will lower thresholds to get higher sensitivity for earlier warnings of developing degradation mode. However, with variable performance assets and especially with variability in ambient vibration and exogenous noise, lowering the thresholds results in spurious alarms that have no significance at all. In most industries, false alarms from spurious trips of thresholds are extremely costly, because the false alarms result in taking assets out of service unnecessarily. As a response, monitoring system designers raise the alarm thresholds on peak frequencies, and then if there is a new degradation mode over the degradation is severely underway, the machine is dead before any alarm is issued [11,12]. For rotating machinery with dynamic workloads, variable-speed performance, and mounted into structures containing other dynamically varying vibration sources, thresholds on gross vibrational amplitudes are very inefficient for detecting the early onset of degradation. The inefficiencies are a consequence of the threshold boundaries. Threshold alarm boundaries must be set higher than the highest peak for the component at its highest load, highest RPM setting, and when the ambient vibration levels are highest. This significantly lowers the "early warning" potential for prognostics, because thresholds on vibration amplitudes can be very inefficient when components are not operating at peak load/performance conditions.

One solution to the threshold inefficiencies is to introduce an ML model for predictive monitoring of the vibration and acoustic signals. A simple univariate model could be used but multivariate models are significantly more accurate. Unfortunately, vibration signals are generally univariate. A secondary hindrance is that vibration measurements are inherently high frequency, sample rates for vibrations are in the kHz range. The glut of data generated by these sensors quickly becomes unmanageable for storage much less analysis by algorithms more complex than monitoring basic statistical moments changes in the time series. Therefore, monitoring vibration becomes untenable [12–15]. The Vibration and Acoustic Resonance Spectrometry (VARS) processing is a solution that allows for lossless dimension reduction of the signals and produces a multivariate dataset from one vibration sensor that can be consumed by an ML model. The ML signal-processing algorithm called the Multivariate State Estimation Technique (MSET) is a non-linear regression-based technique. It is best suited for anomaly detection in time series data. It has been widely adopted in many business-critical industrial for more than 20 years. It has been redeveloped it in the recent years to make it scalable to large collections of telemetry sensors, for big prognostic use cases. There are several reasons for such widespread adoption. The MSET has significant advantages regarding great accuracy and low false-alarm and missed-alarm rate. They are crucial especially in the field of anomaly detection, false alarm leads to unnecessary asset shutdown and missed alarms can lead to catastrophic failures, both are very costly. Furthermore, the overhead compute cost of MSET is much lower compared to the other prognostic algorithms such as neural networks and support vector machines, because of the its deterministic mathematical model, this is important for streaming prognostics which involves tens of thousands of sensors [10,13,16–19].

The VARS process that is tested in this paper is upstream of MSET. As an overview, VARS transforms a single univariate vibration or acoustic signal into 20 correlated time series signals that is predictive of an asset's unique operational frequency signature. First, during the measurement phase, a deterministic periodic loading pattern, that covers the entire range of the asset operation, is introduced to excite the appropriate system dynamics in the frequency domain. The second stage is a preprocessing algorithm and is the cornerstone of VARS. The second stage is composed of several substages whereby the accelerometer and/or acoustic measurements are manipulated with a frequency-domain to time-domain to frequency-domain double transformation. The double transformation ostensibly reduces the dimension of the data without loss of prognostic information. In the final stage the output from VARS is input into an ML algorithm generating a model for subsequent monitoring of new measurements. The ML monitoring informs condition-based decisions about the state of the mechanical, electromechanical, or thermal-hydraulic system.

The rest of this paper is structured as follows. Section 2 will discuss the specifics of the VRS methodology along with some design choices and algorithmic structures. Section 3 will briefly discuss MSET2 and the process by ML monitoring of telemetry occurs. Section 4 will discuss our current experimental setup. Section 5 will explicate our results for the current experiment with a conclusion in Section 6.

2. DETECTION METHODOLOGY:

2.1 Principals of the Method

The VARS, when combined with other auxiliary supporting computational agents, is capable of robust prognostics for all types of mechanical, electromechanical, and thermal-hydraulic flow systems. The algorithm autonomously discriminates the multi-bin operation frequency signatures for predictive-based health monitoring of an asset. Predicated on solving two major hindrances to vibration predictive monitoring, high frequency sample rates, and low sensitivity to subtle degradation, the algorithm solves these inadequacies in a multi-stage analytical framework. First, during the measurement phase, an oscillatory loading pattern, that covers the entire range of the asset operation, is introduced to excite the appropriate system dynamics in the frequency domain. The second stage is a preprocessing algorithm and is the cornerstone of VARS. The second stage is composed of several substages whereby the accelerometer or acoustic measurements are manipulated with a frequency-domain to time-domain to frequency-domain double transformation. The double transformation ostensibly reduces the dimension of the data without loss of prognostic information. In the final stage the output from VARS, several time series signals, is input into an ML algorithm generating a model for subsequent monitoring of new measurements. The ML monitoring informs predictive-based decisions about the state of the mechanical, electromechanical, or thermal-hydraulic system.

2.2 Vibration Measurement Training Regimen

When utilizing ML to monitor an asset it is a requirement for the training data to encompass the entire range of the system operation. For example, a model, monitoring a car, trained from data containing only surface street driving measurements will generate alarms if the car were to begin highway driving. Additionally, it is common for mechanical assets to operate in small subset of the spectrum of operation for long periods. To continue with the car as an example assume it operates mostly in the city. The driving will be on the streets, oscillating between idle and approximately 40 miles per hour, with either short spurts of regular freeway driving or long-range driving on the freeway but with much less frequency. In the first scenario, it will take a long time to log enough freeway training data to generate reliable predictions. In the second scenario, one long trip may provide enough freeway measurements for a valid training set but with much less opportunity to do so. Furthermore, most drivers never reach peak or maximum performance in a car regardless how long and how often the car is driven. Therefore, it is necessary to incite the entire operational range during the initial stage of measurements to guarantee sufficient training data to generate a reliable model.

Another concern that is specific to vibration measurements is the magnitude of the stochastic measurement noise. Tracking vibration modes requires high resolution and dense sampling to capture the true dynamics of the system, which are otherwise opaque, allowing for more opportunity for noisy measurements. Noise can also stem from how the vibration of a system is being measured often, vibrations are tracked globally and not through individual components in the system. This is especially true for acoustic measurements. Tracking global measurements as opposed to individual components of the system are generally more pragmatic for example, storing and processing data from all the components in a system is prohibitively expensive or access to the components may unfeasible. Moreover, global measurements track multiple components in parallel and the emergent behavior of the system is more indicative of reality. Unfortunately, tracking system behavior generates noisy measurements because it is composed of multiple components that vibrate and resonate with their own unique frequencies.

To mitigate noise and guarantee adequate training data the monitored asset runs through a short oscillatory loading pattern for a nominal amount of time. The workload must cycle through all known, valid, operational states of the system. The pattern can be sophisticated, if the asset has programmable workload capability, or can be extremely trivial such as simply toggling On/Off, or Idle/Max, several times in a "training" window. This training measurement regime ensures full operational classification and aids in processing downstream.

2.3 VARS Preprocessing

The approach for vibration analysis introduced in this paper is comprised of a frequency-domain to time-domain to frequency-domain double transformation. Resulting in an optimal set of a finite number of narrow-band frequency bins. This creates a set time series signals. The bins are sampled at rate which matches other sensor telemetry for the assets under surveillance, e.g. once every 2 secs or every 5 secs. Ideally, the fastest sampling rate for other telemetry metrics (e.g. temperatures, voltages, currents, RPMs, etc.) is selected to sample the narrow-band frequency bin time series thereby instituting real-time vibration monitoring. The bins are then averaged and ranked resulting in a set of time series which are subsequently consumed by a machine learning algorithm, along with, additional telemetry, to monitor the condition of rotating machinery, or and oil pump, or other mechanical systems.

Once the raw measurements are recorded, there is an initial transformation into the frequency domain by utilizing a fast Fourier transform (FFT) and the PSD is calculated. The window or sample size for the FFT is determined by the fastest sample rate of the remaining telemetry monitoring any given mechanical system, this is the first stage in dimension reduction. The results are a set of time series that records the change in frequency across the entire frequency spectrum, up to the Nyquist frequency. While there are some distinctly excited frequencies that are representative of the operating frequency of the system, they lie in narrow bands of the entire sample spectrum. To isolate the operational signature narrow bands of frequencies are grouped, or "binned," and undergo a process for reducing noise within each group and ranking.
To identify the more representative frequencies we split the frequency spectrum into several frequency bins. Subsequently, a technique to generate a time series signal that is a correlated average of the frequencies in the bin is applied. The averaging reduces the problem size and increases the signal to noise ratio (SNR) of the bin.

The first step is to decide the number of frequency bins to use for subdividing the spectrum. The number of total bins has been determined to be 100. The value been determined empirically through related research [20,21]. A reduction or increase in this value is dependent on the context but 100 is at the optimal intersection between dimension reduction and robust frequency characterization of an asset.

The number of frequencies in each bin will also influence the results. If each bin is of equal length, it is likely that a portion of the spectrum that is either minimally active or superfluous will have a disproportionate impact on the results. The application of weighting system based on the Root Mean Square (RMS) accounts for inequity in the measurement spectrum. Where A is the discrete fast Fourier transform (DFFT) of the vibration signal. The RMS was natural choice as it is a useful metric for determining the overall energy in each portion of the spectrum. Weighting the segments of the spectrum with more energy in this manner reduces the bin sizes, which proportionally increases the number of bins in those ranges.

$$RMS = \sqrt{\frac{A_0^2}{2} + \sum_{i=1}^{k-1} A_i^2 + \frac{A_k^2}{2}} \quad (1)$$

After weighing and grouping the spectrum the data reduction begins in kind with a correlated averaging of the frequencies in each bin but many of the bins, even after grouping, are just random noise. Bins with more correlated frequencies correspond to the load conditions and the operation frequency of the asset. Therefore, we apply an algorithm that determines the most correlated signals in a bin and averages these signals before ranking them. All the signals in the bin go through a correlation matrix assessment. The assessment sums the correlation coefficients of the signals, sorting from lowest to highest. The mean of the signals in the top percentile yields a single time series signal for each frequency bin. The use of percentile, as opposed to correlation coefficient, as the metric generalizes the approach to be adaptive to any size frequency bin with any SNR distribution.

To illustrate the motivations behind the correlation ranking average an example of one is the 100 frequency bins is shown in Fig. 1, below. We can see the variance inside the bin resulting in a noisy set measurement which, is makes predictions and uniform reproduction difficult. Instead, a subset of signals inside the frequency bin with the best inter-correlations are averaged, filtering out the noise, yielding a high noise ratio signal (green) and revealing the dynamic pattern of the workload.

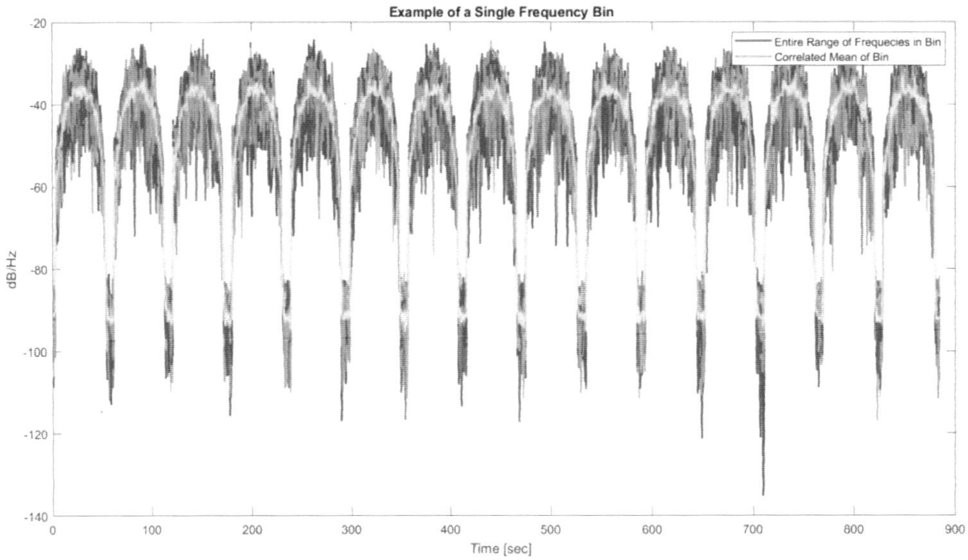

Fig. 1. The example frequency bin consists of the fine grain frequencies measurements between two frequency ranges in the spectrum.

After the binning step, the frequency bins are sorted. The signals most representative of the operational frequencies of the system in the spectrogram are chosen while the remainder are discarded. The loading pattern conducted during the measurement step enhances the sorting. To determine the strongest frequency components the resultant 100 time series from the binning step are transformed into the frequency domain again. The bins are sorted by the magnitude of the PSD at the period of the loading pattern. After the sorting has occurred the number of bins that are discarded is dependent on the sensitivity requirements of the system, 20% of the number of discrete frequency bins is recommended for selecting the top signals based on the height of the peak in the frequency domain. As with the 100 bins the top 20% have been determined, through empirical use, to be the most representative of the frequency behavior of most assets [20,21].

3. MSET - ML Based Monitoring:

The MSET has two primary phases: training and inferencing. The model is trained with operating-frequency-data from a system. Here the system does not have to be a brand-new asset but is an asset that is certified to have no degradation, and with this model, MSET estimates are produced for new data. After that, the pairwise residuals between MSET estimates and actual values are computed and sent to an anomaly detection module called Sequential Probability Ratio Test (SPRT) [17–19]. The SPRT performs two statistical hypothesis tests, where the mean and variance shift between the reference distribution and degraded distribution are quantified to identify the anomalies. SPRT allows users to specify both false alarm rate and missed alarm rate, this avoids the conventional trade-offs between the low false-alarm rate and low missed alarm rate.

The training for MSET is composed measurements taken from a healthy system. The measurements are consistent of highly correlated features from a mechanical system, such as RPMs, Voltages, and Currents. In this case, the training will consist of the 20 frequencies identified during a period of healthy operation.
During the monitoring phase, the measurements will be continued after the system is altered. The alteration in this case will consist of faulty gears in the gear box. The frequency bins that are monitored for the faulty gear measurements are identical to the 20 bins determined during the binning and sorting phase while the system is healthy. The motivation for doing so is that often when damage occurs in a system a shift in operational frequency occurs. For example, if a gear tooth in a gearbox chips the motor power and RPMs will likely increase to compensate for the damage, resulting in higher frequency vibrations [22]. The frequency behavior of the healthy state will be most indicative of normality.

4. EXPERIMENTAL SETUP:

The experimental testing is performed on a drivetrain dynamics simulator designed to simulate industrial drivetrains, which consists of the variable speed driver, gearbox, magnetic brake, and kinds of sensors. The variable speed driver is a 3-phase, 3-hp motor, and pre-wired self-aligning mounting system for easy installation, which is driven with a multi-featured front panel programmable controller. The maximum speed of the motor is 5000 rpm. The gearbox consists of a two-stage and 2.5 transmission ratio gear system with spur and helical gear. The magnetic brake provides a 1.5 to 32 lb-ft capacity heavy duty.

Fig.2. Experimental setup.

The entire experimental setup is depicted in Fig.2. During the measurement, the driver speed is a periodical workload, which ramped up and down from 0 Hz to 60 Hz. The programmable magnetic brake provides the exact load setting. The tri-accelerometer is placed on top of the gear housing and the microphone is placed in the upper front of the gearbox and avoids touching the gearbox. The missing tooth of the helical and spur gear is used as the gear fault type. The measurement process was repeated 5 times at a sampling frequency of 20.48 kHz, each lasting 900s.

5. RESULTS AND DISCUSSION:

To guarantee the full range of operation the of the drive train system (DDS) were included in the training data and to more easily identify the frequency bins that are unique to the system oscillatory workload was introduced into the measurements. The SpectraQuest DDS has the capacity for programming the rotational frequency of the motor so that the system can cycle through all operational modes. Additionally, the power load can also be manipulated but was kept constant for the initial research. The oscillatory loading pattern was designed to be sinusoidal and illustrated in Fig. 3.

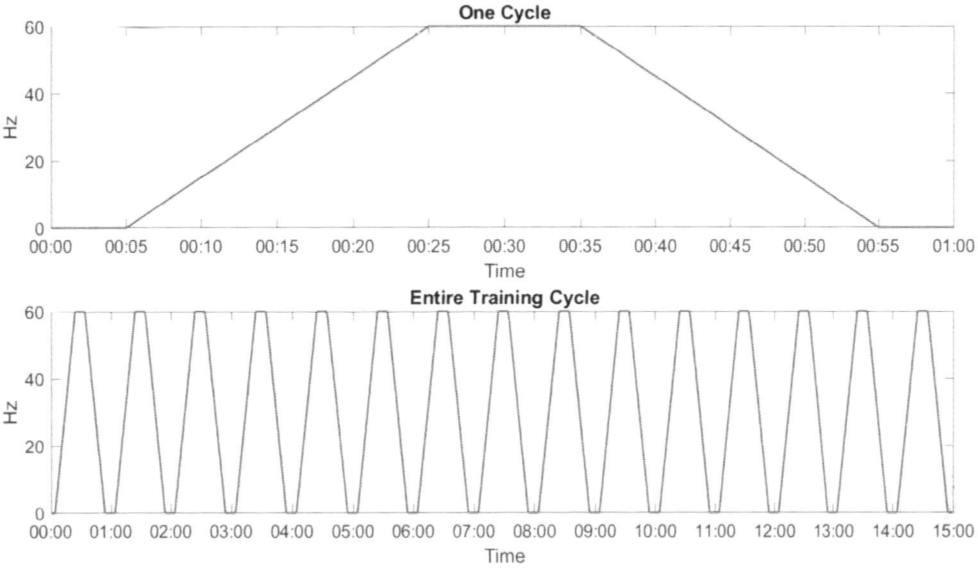

Fig. 3. The oscillatory workload implemented during the experimental measurements. One period is equal to one minute. To generate the wave the motor frequency and system begins from static equilibrium cycles up to 60 Hz, the maximum frequency, for 25 seconds, remains at maximum frequency for 10 seconds, subsequently ramping back down to the static state. The cycle was repeated over a 15-minute period for both healthy and damaged states.

5.1 VARS and Predictive Monitoring

The first step in the VARS process is to convert the measurements into the frequency domain. To accomplish the transform the signals are windowed to the intended sample rate, Fig 4. For simplicity, the sample window chosen for these experiments is one second, indicated by the vertical red line. The measurements last for 900 seconds but only 7 seconds are shown, to make the window size more visually legible.

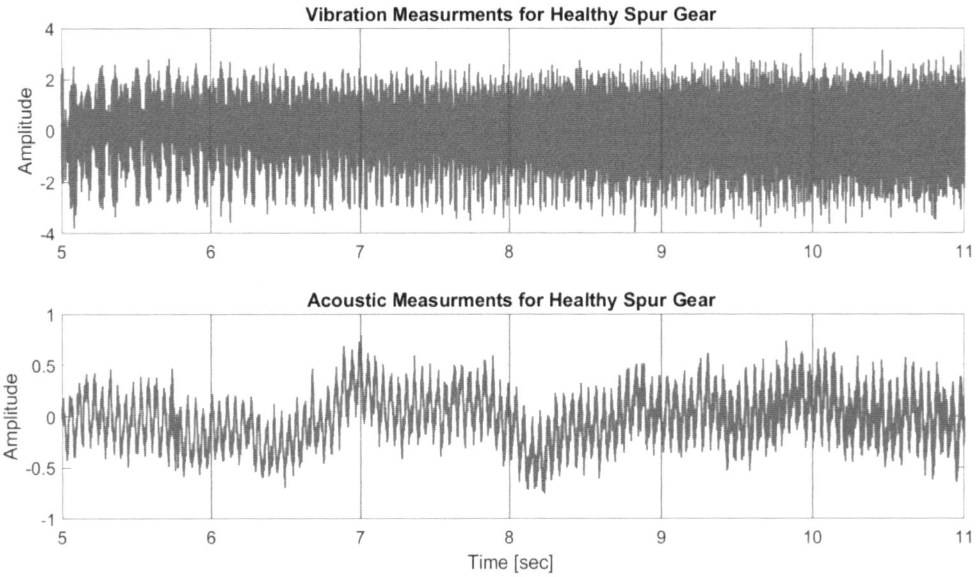

Fig. 4. Vibrations (top) and acoustic (bottom) measurements for windowing.

Each window is then transformed into the frequency domain through an FFT. Subsequently the PSD is calculated resulting in a granular set of a time series which track the frequency change across the entire sample spectrum, for each state of the fan. A moving Hann window is applied to minimize edge artifacts in the frequency domain. In Fig. 5 we illustrate this transformation for both the vibration and acoustic signals for the one healthy state run.

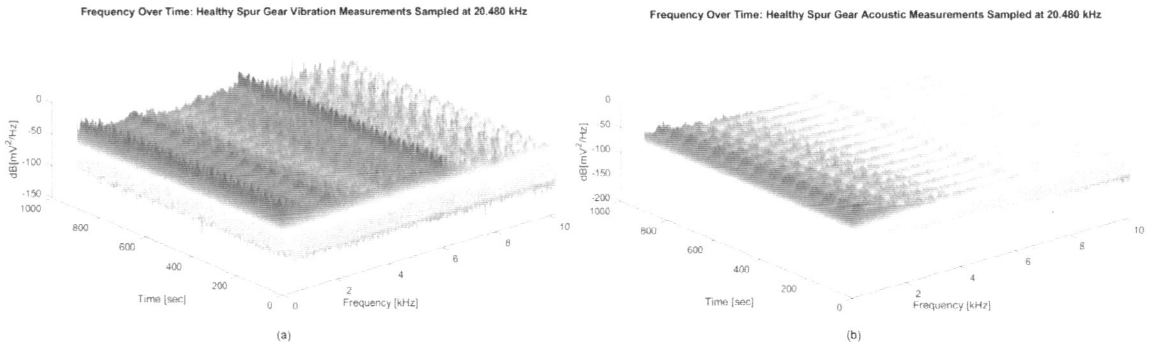

Fig. 5. The changing frequency over the period of measurement for the healthy state of the gear.

The dynamics of the oscillatory workload are visually apparent in both the acoustic and vibration measurements. While the vibration measurements, in Fig. 5 (a), are relatively consistent with three distinct ranges of peak activity occurring in the first 1000 Hz, around the 2kHz, and again around the 6kHz range, the acoustic measurements, in Fig. 5 (b), are frontload. Most of the amplitude in the spectrum is in the first 1000 Hz. The binning for both signals is weighted accordingly to account for the distribution of the active frequencies.

To determine the size of each frequency bin and therefore the weighting of each segment of the spectrum we utilize the RMS metric (eq. 1) to ascertain the energy content in each part of the spectrum. First, a PSD of the entire spectrum for the healthy state measurements, as seen in Fig. 6 (a), is determined. Next the cumulative RMS is calculated, Fig. 6 (b), across the entire frequency spectrum to assess the change in slope, which corresponds to an increase in the energy content of the spectrum, for subdivisions of these spectrum. Then these subdivisions are weighted by the percentage of the RMS they contain, see Fig. 6 (c) and (d). For instance, in Fig. 6 (c) the percentage of the RMS that is contained within the first 1000 Hz of the vibration measurements is approximately 40% which will translate to 40 bins, out of the 100 total bins, being dedicated to the initial 1kHz.

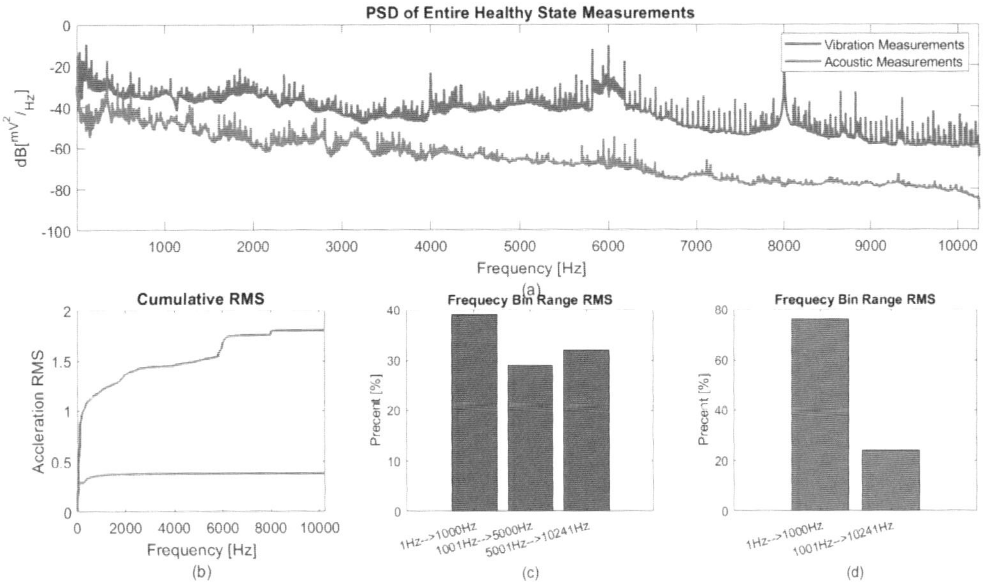

Fig. 6. The sequence for determining the frequency bin weights. (a) PSD of entire healthy state measurements. (b) Cumulative RMS. (c) and (d)Percentage of RMS for frequency bin range.

As aforementioned one of the main motivations behind the oscillatory workload is to assist in the ranking of the frequency bins. In this experiment, the periodicity of the workload is 1 min, or 60 seconds. As such, the largest magnitude of the PSD for the germane frequency bins will be at 60 seconds. As shown in Fig. 7 the top 2 frequency bins, corresponding to both the vibration (Fig. 7 (a)) and acoustic (Fig. 7 (b)) measurements for the healthy spur gear.

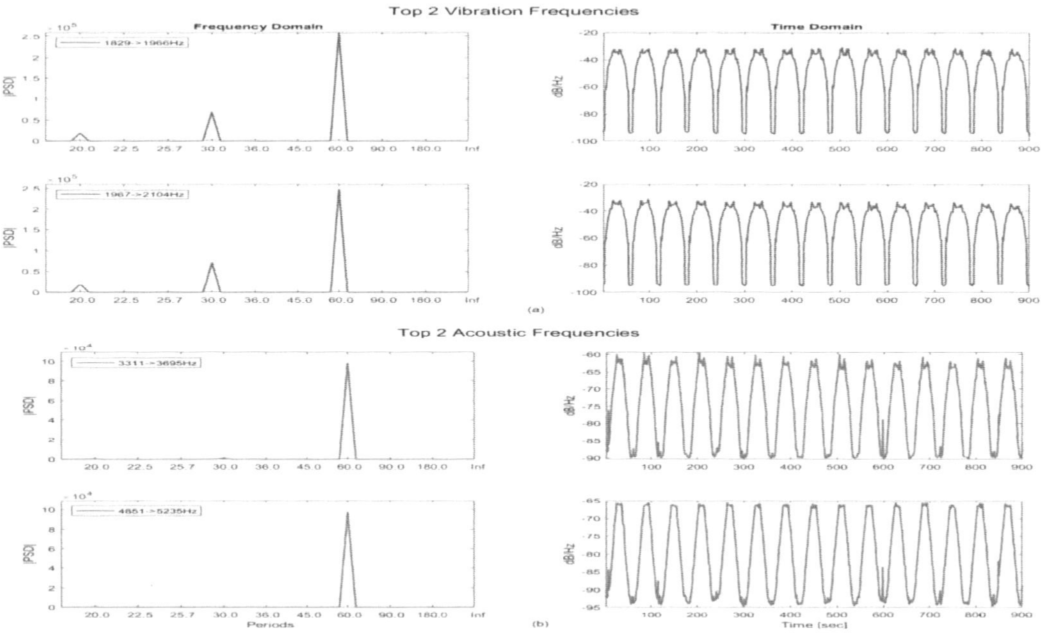

Fig. 7. Illustration of the top 2 frequency bins after sorting occurs when the periodic load is introduced.

With the VARS process and the periodic workload identifying the frequency bins that most closely couple asset mechanical performance with output vibrational measurements. These signals are the optimum narrow-band signals to include in the subsequent time-series machine learning monitoring.

The final step is to apply MSET to the signals to generate a model to conduct predictive maintenance on incoming measurements. Generally, the 20 frequencies determined from the healthy state measurements from an asset will be the training data for MSET. This model is used to monitor new data for the same asset as it changes over time. In this instance there will be an assessment of the VARS process to illustrate the value, in combination with MSET, for predictive monitoring. In this instance we evaluate the process with the fan data.

5.2 Healthy State Monitoring

Initially we were interested in determining the prominence of false alarms in data that were healthy by modeling and monitoring multiple sets of healthy measurements. To accomplish this task MSET is first trained on one set of healthy state measurements and monitors another. The frequencies chosen for the monitored measurements are the same frequency indexes chosen for the training. To illustrate how MSET would operate we present an example signal, presented in Fig. 8. In this example one set of healthy state acoustic measurements is compared to another set of acoustic measurements. In the top plot the generated MSET estimates, in green, are compared with the actual measurements, represented in red. Subsequently the residual is calculated, second subplot, to determine the difference between the estimates and the measurements. The SPRT test is then applied, to determine at what point in time, if any, there is a deviation in the distribution of the residuals. Lastly, the results from the SPRT test, bottom subplot, are presented as a binary indicator for the SPRT alerts where 0 means healthy and 1 indicates damaged. As expected, there is no indication of any degradation. This process was repeated with all permutations of measurements for both the vibration and acoustic data. Both types of measurements were tested along with the multiple gear types for the gearbox.

5.3 Fault Detection

The final step in the process is to utilize MSET for predictive monitoring as an asset that may begin to degrade, in this case there were several faulty gear replacements that were introduced to the DDS. To illustrate how MSET would operate as a real time monitoring system we concatenated the healthy measurements with the faulty gear states. First, the 20 frequencies from the healthy state measurements are determined from VARS. We then utilize the same 20 frequency bin indexes determined from healthy state to pick the 20 bins for the faulty state and concatenate the healthy measurements with the faulty ones. The concatenated signals are intended to simulate a system that suddenly begins to degrade. The resulting dataset is 30 min of measurements across 20 frequency bins. The MSET is trained on the first 850 seconds then monitors the remaining 950 seconds. One minute of the healthy measurements remain in the testing phase to illustrate the differentiation capacity of the model.

The monitoring stage is presented in Fig 9 with one example signal from the set of vibration measurements where the helical gear replacement was cracked. As with the healthy state example, there are three images, the signal measurements and estimates, top, the residuals between the two, second subplot, and the results from the SPRT test, bottom. Conversely, to the previous example there is degradation in the signals. In this instance, the unbalanced measurements are introduced to the testing phase at the 900th second. The MSET is able to detect this sudden change in operation, as indicated by the SPRT alarms (red) in the bottom subplot.

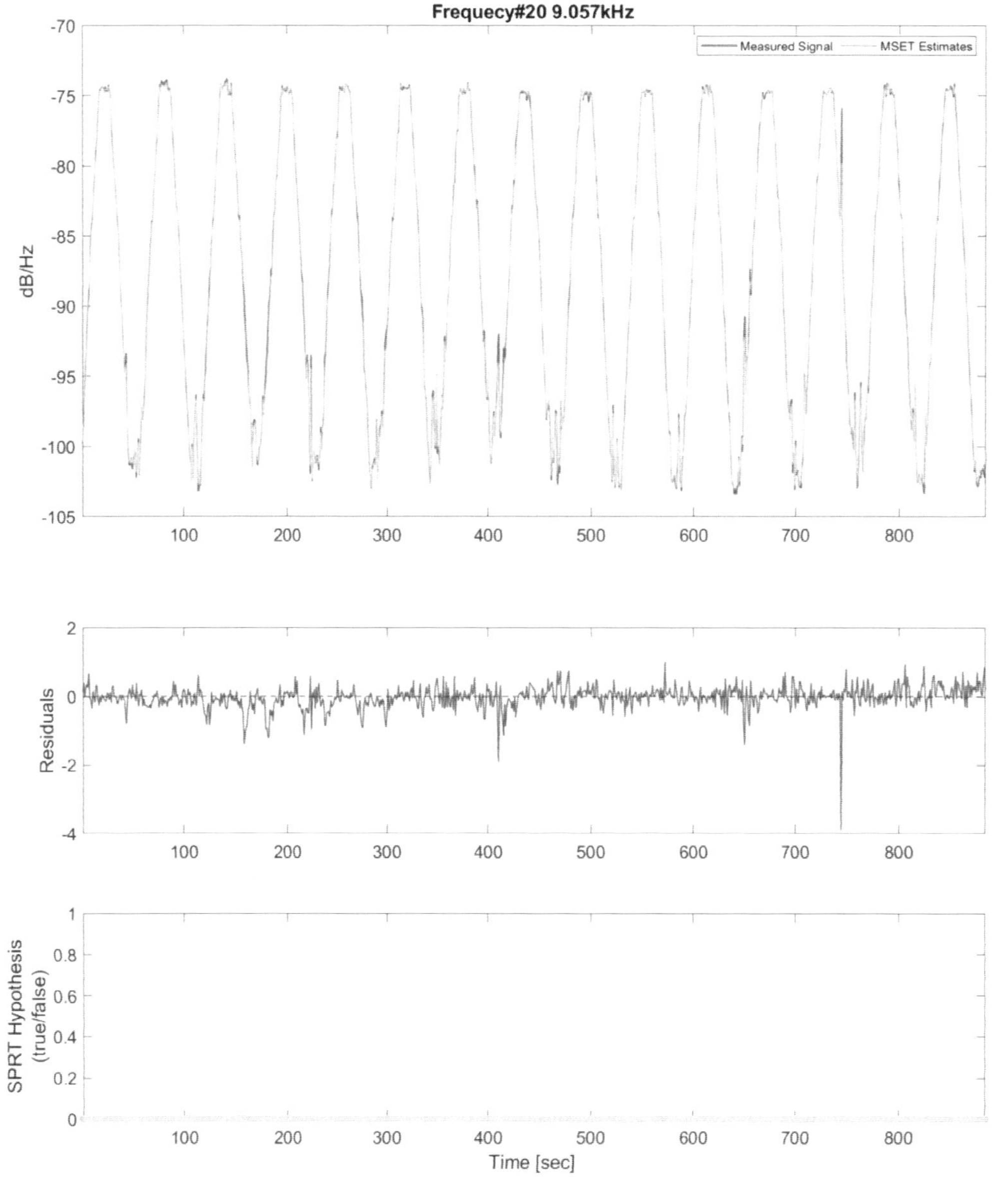

Fig. 8. Visualization of MSET monitoring a healthy gear at the frequency of 9.057 kHz.

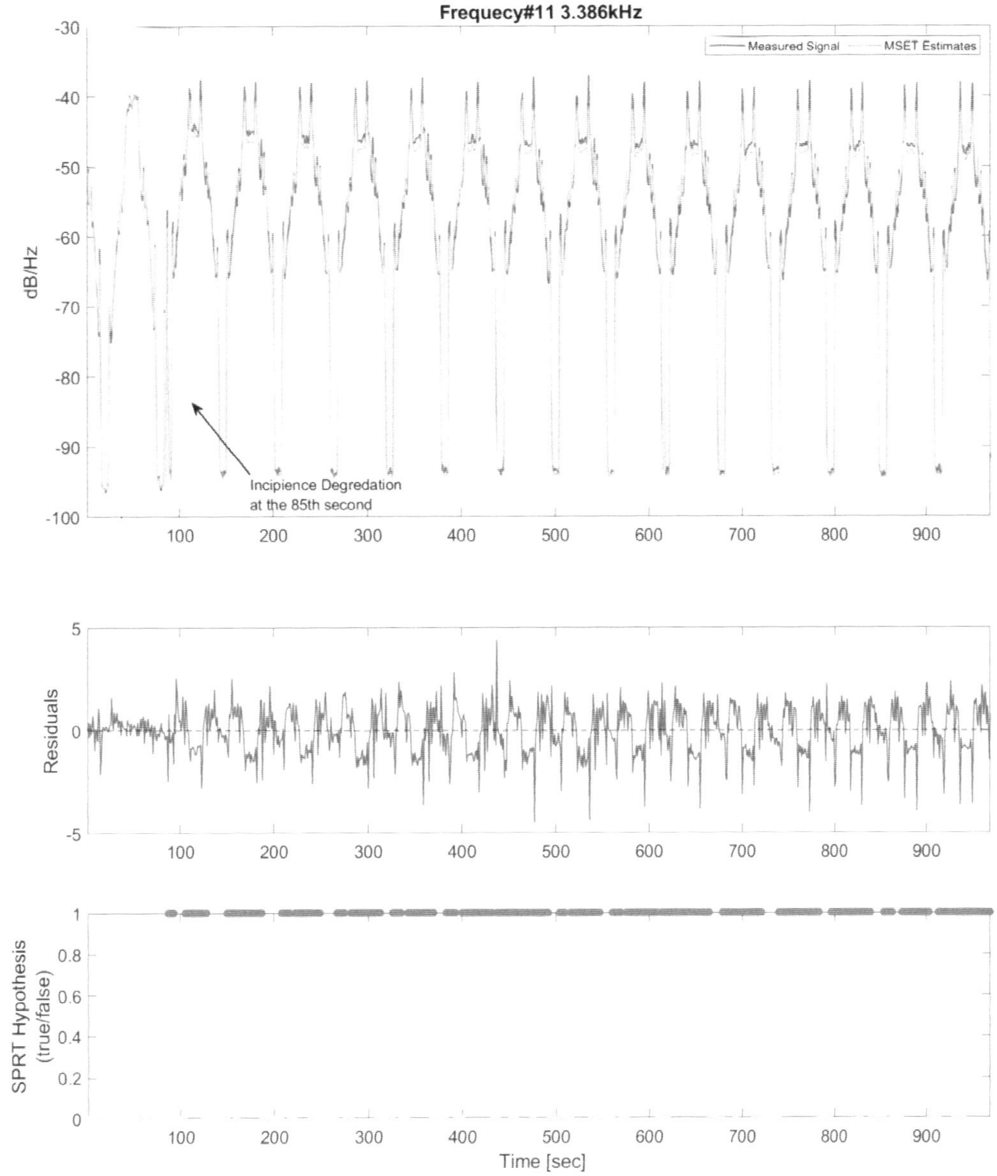

Fig. 9. Visualization of MSET monitoring of the fault gear at the frequency of 3.386 kHz.

As with the healthy state measurements a methodical, iterative, evaluation whereby all possible permutations of training and monitoring of the balanced and unbalanced measurements was conducted. In all instances, for both acoustic and vibrations signals we can detect the faults currently researched for both types of gears. Table 1 below contains an overview of the gear types and the fault types VARS and MSET were able to detect.

Table 1. Summary of the faults that have been detected in the DDS.

Gear Type	Vibration		Acoustic	
	Missing Tooth	Cracked	Missing Tooth	Cracked
Spur Gear	Yes	N/A	Yes	N/A
Helical Gear	Yes	Yes	Yes	Yes

6. CONCLUSIONS:

The utilization of the VARS preprocessing algorithm also allows for automated predictive monitoring. The process transforms a noisy univariate signal, into 20 highly correlated signals that can be used in a multivariate ML model, generated either on its own or in combination with other SCADA telemetry. When VARS is placed upstream of MSET the real time condition monitoring becomes more sensitive to the current condition and the onset of degradation or mechanical assists than is currently available industry.

Future research that continues to use the SpectraQuest DDS is underway. The research will begin several other fault types pertaining to planetary gear, other faulty states for the spur gear and the helical Gear, and faulty bearings. Another test will increase in the complexity of the loading patterns will be implemented by oscillating the power input for the DDS to test more modes of operations. Moreover, additional telemetry will also be introduced into the model.

ACKNOWLEDGEMENTS

This research is supported by a grant from the Oracle Corporation.

REFERENCES

[1] Niewczas, P., Dziuda, L., Fusiek, G., and McDonald, J.R., "Design and evaluation of a preprototype hybrid fiber-optic voltage sensor for a remotely interrogated condition monitoring system," *IEEE Transactions on Instrumentation and Measurement* 54(4):1560–1564, 2005.

[2] Niewczas, P., Fusiek, G., and McDonald, J.R., "Dynamic capabilities of the hybrid fiber-optic voltage and current sensors," *SENSORS, 2006 IEEE*, IEEE: 295–298, 2006.

[3] Rabbi, S.F., Rahman, M.A., and Butt, S.D., "Modeling and operation of an interior permanent magnet motor drive for electric submersible pumps," *2014 Oceans-St. John's*, IEEE: 1–5, 2014.

[4] Silva, P.A.S. da, Costa, C.T. da, and Barreiros, J.A.L., "Intelligent Analysis Program Applied to Production Logs in Oil and Gas Wells," *IEEE Latin America Transactions* 4(5):353–358, 2006.

[5] Shanbr, S., Elasha, F., Elforjani, M., and Teixeira, J.A., "Bearing fault detection within wind turbine gearbox," *2017 International Conference on Sensing, Diagnostics, Prognostics, and Control (SDPC)*, IEEE: 565–570, 2017.

[6] Yang, J., Zhao, L., Lang, Z.-Q., and Zhang, Y., "Wind Turbine Blade Condition Monitoring and Damage Detection by Image Based Method and Frequency-Based Analysis," *2018 10th International Conference on Modelling, Identification and Control (ICMIC)*, IEEE: 1–6, 2018.
[7] Chikuruwo, M.N.H., Maregedze, L., and Garikayi, T., "Design of an automated vibration monitoring system for condition based maintenance of a lathe machine (Case study)," *2016 international conference on system reliability and science (ICSRS)*, IEEE: 60–63, 2016.
[8] Seo, J., "A practical scheme for vibration signal measurement-based power transformer on-load tap changer condition monitoring," *2018 Condition Monitoring and Diagnosis (CMD)*, IEEE: 1–4, 2018.
[9] Rikardo, S.A., Bambang, C.B., Sumaryadi, C., Yulian, T.D., Arief, S.E., and Kharil, S.F.I., "Vibration monitoring on power transformer," *2008 International Conference on Condition Monitoring and Diagnosis*, IEEE: 1015–1016, 2008.
[10] Pavithra, R. and Ramachandran, P., "An Overview Of Predictive Maintenance For Industrial Machine Using Vibration Analysis," *2021 Innovations in Power and Advanced Computing Technologies (i-PACT)* 1–7, 2021.
[11] Minaiepour, H., Sarabeigi, H., and Heydari, E., "Surveying compressor C-2501 vibration problem by analysis of vibration and its analytical report," *2012 IEEE International Conference on Condition Monitoring and Diagnosis*, IEEE: 706–709, 2012.
[12] Mo, Y.-C., Su, K.-Y., Kang, W.-B., Chen, L.-B., Chang, W.-J., and Liu, Y.-H., "An FFT-based high-speed spindle monitoring system for analyzing vibrations," *2017 Eleventh International Conference on Sensing Technology (ICST)*, IEEE: 1–4, 2017.
[13] Gross, K.C. and Lu, W., "Early detection of signal and process anomalies in enterprise computing systems.," *ICMLA*, 204–210, 2002.
[14] Durgam, S., Bawankule, L.N., and Khindkar, P.S., "Prediction of Fault Detection Based on Vibration Analysis for Motor Applications," *2021 4th Biennial International Conference on Nascent Technologies in Engineering (ICNTE)*, IEEE: 1–5, 2021.
[15] Yuan, X., Gao, Z., Tang, K., Zeng, M., and Wang, Y., "Wind turbine gearbox condition monitoring system based on vibration signal," *2015 12th IEEE International Conference on Electronic Measurement & Instruments (ICEMI)*, IEEE: 159–163, 2015.
[16] Sharma, R. and Pandey, N., "A neural network model for electric submersible pump surveillance," *2016 International Conference on Communication and Signal Processing (ICCSP)*, 2083–2088, 2016, doi:10.1109/ICCSP.2016.7754545.
[17] Gross, K.C., Baclawski, K., Chan, E.S., Gawlick, D., Ghoneimy, A., Liu, Z.H., and Zhang, X., "MSET Prognostics for Operator Decision Aid for Human-in-the-Loop Supervisory Control Applications," *2016 IEEE International Multi-Disciplinary Conference on Cognitive Methods inSituation Awareness and Decision Support (CogSIMA)*, San Diego, CA, 2016.
[18] Gross, K.C., McMaster, S., Porter, A., Urmanov, A., and Votta, L.G., Proactive system maintenance using software telemetry, 2003.
[19] Gross, K.C. and Humenik, K.E., "Sequential probability ratio test for nuclear plant component surveillance," *Nuclear Technology* 93(2):131–137, 1991.

[20] Wetherbee, E.R., Wang, G.C., Gross, K.C., Dayringer, M., Lewis, A., and Gerdes, M.T., "Counterfeit device detection using EMI fingerprints," Google Patents, 2022.

[21] Liu, Y., Gerdes, M.T., Wang, G.C., Gross, K.C., and Balasubramanian, H., "Autonomous discrimination of operation vibration signals," Google Patents, 2023.

[22] Durbhaka, G.K. and Selvaraj, B., "Predictive maintenance for wind turbine diagnostics using vibration signal analysis based on collaborative recommendation approach," *2016 International conference on advances in computing, communications and informatics (ICACCI)*, IEEE: 1839–1842, 2016.

MFPT ANNUAL CONFERENCE
Where Theory meets Practice

MITIGATION OF EXCESSIVE VIBRATION AT MOTOR NDE DUE TO RESONANCE WITH TUNED MASS DAMPER

Meng Hee Lim, Zair Asrar, Kee Quen Lee, and Mohd Salman Leong

Institute of Noise and Vibration
Universiti Teknologi Malaysia Kuala Lumpur
Jalan Sultan Yahya Petra
54100 Kuala Lumpur, Malaysia
mhlim.kl@utm.my

ABSTRACT:
This paper presents a case study for fault diagnosis and mitigation of excessive motor vibrations caused by floor-induced resonance. Vertical motors in a water treatment plant had experienced excessive vibration levels (> 12 mm/s) at its non-drive end (NDE) location. A comprehensive vibration investigation was conducted to map-out the vibrations of the entire machine train which includes pipes, coupling, motors, pedestal, plinth and floor slab of the pump house. Vibration investigation found that the root cause of the excessive motor vibration was caused by the resonance of the floor slab as the natural frequency of the floor slab (20 Hz) coincided with the operating frequency (25 Hz) of the motors. As such, the first remedy action undertaken at that time was the reinforcement of the motor pedestal with additional ribs in order to shift the primary natural frequency of the motor-pedestal system to be away from the motor operating frequency. However, this attempt failed to subside the excessive motor vibration as this method did not result in significant shift in the natural frequency of the motor-pedestal-floor system. The second attempt was conceived to design a tuned-mass damper (TMD) for the motors. Finite Element Analysis was used to design a custom made TMD to be attached at the motor NDE locations. After several fine-tuning and modifications, the TMD was installed and had successfully reduced the motor NDE vibration levels for the vertical motors to be within 5 mm/s for all tri-axes measurements and thus enables the motors for long term unrestricted operation in accordance with the ISO 10816-3 Standard.

KEY WORDS: Floor Resonance; Motor NDE; Tuned-Mass Damper

INTRODUCTION:

This paper presents a comprehensive case study on vibration investigation conducted at a water treatment plant in Malaysia. The plant was equipped with six centrifugal water pumps, each driven by a vertical motor. The motors were securely bolted to pedestals, which were further fastened to the plinth on the building's floor slab (refer to Fig. 1). Notably, the motors were installed one floor level above the water pumps. A long composite shaft was employed to connect the vertical motors to their respective water pumps, forming an integral part of the system.

Figure 1: Vertical motors that drive pumps in a water treatment plant

Of particular concern were two motor units in the plant. Motor unit 6 was situated adjacent to the staircase opening on the floor and exhibited exceptionally high vibration levels at its Non-Drive-End (NDE), exceeding 18 mm/s. Similarly, motor unit 5, positioned in the middle of the floor, also experienced elevated vibration levels at its motor NDE location, surpassing 12 mm/s. The excessive vibration of the pumps raised concerns among plant personnel, as the possibility of premature machine failures became a potential risk if the issue persisted.

To identify the root cause of the problem and provide suitable recommendations for remedial actions, comprehensive vibration measurements were conducted for the entire motor-pump units, including the structural floor slabs. This extensive investigation aimed to pinpoint the underlying factors contributing to the excessive vibration and offer effective solutions to address the issue and prevent further complications.

MEASUREMENT STRATEGY AND PROCEDURE:

Vibration measurements were systematically conducted to gain a comprehensive understanding of the characteristics of the elevated motor NDE vibration before implementing any remedial actions on-site. These measurements encompassed the entire machine train, which included the motor-pump system, connected pipes, plinth, and floor slab. To capture accurate data, a multichannel data logger was utilized to record raw time signals from accelerometers placed in three axes. A total of 30 vibration channels were strategically deployed to cover the entire machine train and floor slab, as indicated in Fig. 2.

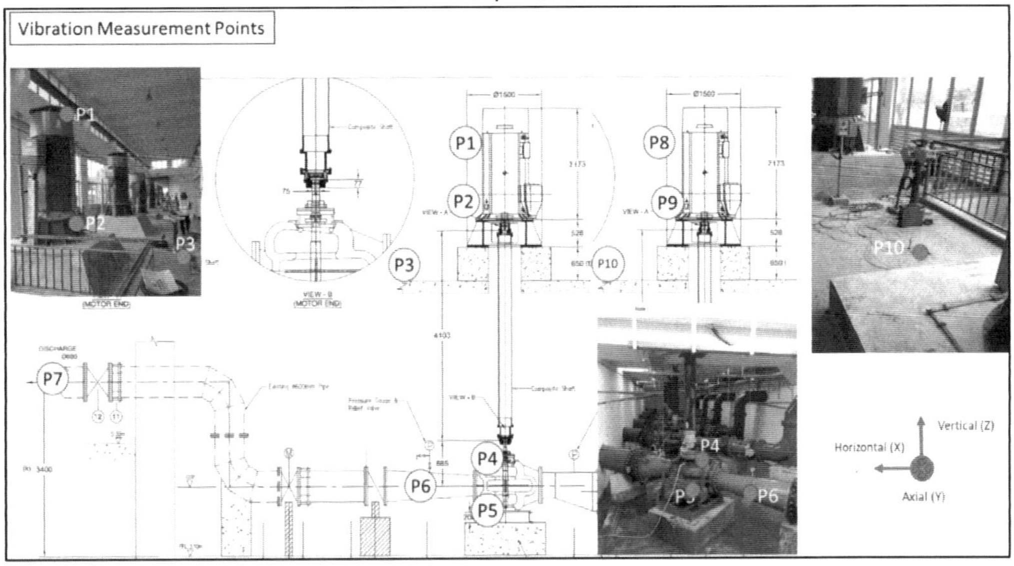

Figure 2: The schematic diagram illustrates a subset of the measurement locations (P1-P10) that were strategically deployed during the measurement exercise.

In addition to the vibration mapping, bump tests analysis was performed on-site using a portable vibration analyzer at critical locations. This approach enabled the identification of the natural frequencies of both the motor-pump system and the structural floor slab. The bump tests were conducted at key locations, including the Non-Drive-End (NDE) and Drive-End (DE) of the motor, discharge pipe, and the floor slab adjacent to the motors. By conducting the bump tests, valuable insights into the dynamic behavior of the system were obtained, allowing for a more comprehensive assessment of the vibration characteristics and the determination of the resonant frequencies associated with the motor-pump system and the floor slab.

RESULTS AND DISCUSSION:

The highest vibration levels for pump units 5 and pump unit 6 were found at the motor NDE location which were measured to be 12 mm/s and 18 mm/s respectively. These vibration levels are considered unacceptable in accordance with the ISO 10816-3 standard. Based on the vibration signal frequency spectrum, the highest contribution was coming from excitation frequency at 1X operating speed of the motor which was at 25 Hz. The dominant direction for high frequency vibration at the motor NDE location was recorded in the y axis (see Fig. 3).

Bump tests were conducted to evaluate the natural frequencies at the NDE of the motor units and the adjacent floor slab. The results revealed that motor unit 5 and 6 exhibited a fundamental natural frequency of 25 Hz, as depicted in Figure 4. This frequency coincided with the operating speed of the motors. Additionally, it was observed that the natural frequencies of the floor slab adjacent to these units also ranged from 20 to 30 Hz, which are close to the natural frequency of the motor NDE. These findings strongly suggested that the high vibration levels observed at the motor NDE of units 5 and 6 were likely a result of resonance induced by the motor's operating speed. The close match between the natural frequencies of the motor and the adjacent floor slab further supported this conclusion.

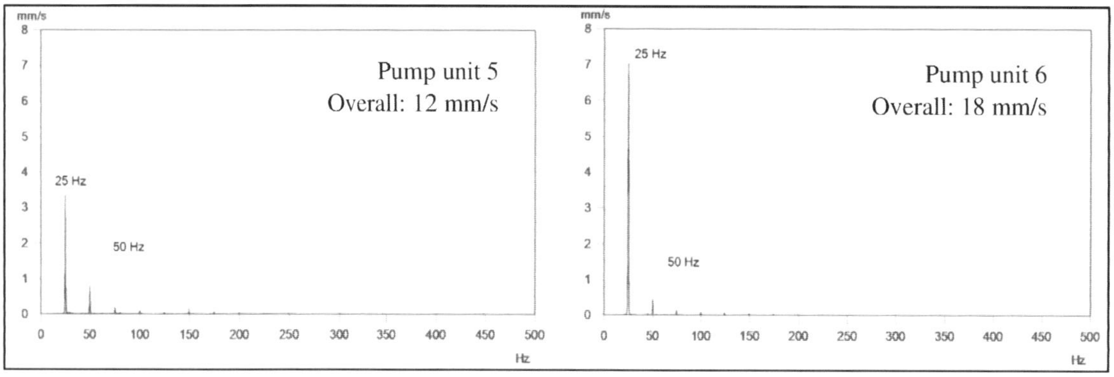

Figure 3: The vibration levels of motor NDE for pump unit 5 and 6.

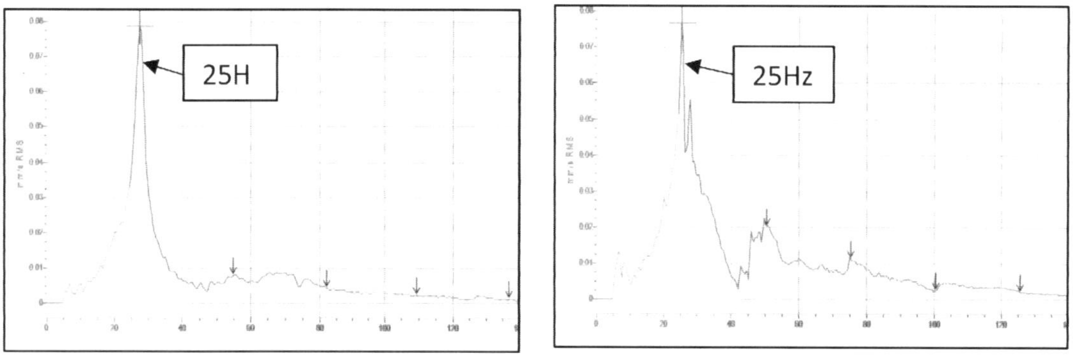

a) Bump test result showed the natural frequency of NDE Motor unit 5, 25Hz

b) Bump test result showed the Natural frequency of NDE Motor unit 6, 25Hz

Figure 4: Natural frequency of motor NDE for a) pump unit 5 and b) Pump unit 6.

A waterfall plot, presented in Figure 5, displays the results of a run-up test conducted at the Non-Drive-End (NDE) of motor unit 5. The plot illustrates a surge in vibration levels observed at this motor, indicating that the elevated vibration levels were indeed a consequence of resonance. Specifically, as the motor speed approached the steady-state operating speed of 25 Hz, the vibration levels experienced a significant increase, further supporting the resonance hypothesis. The waterfall plot provides visual evidence of the relationship between motor speed and vibration levels, highlighting the critical role of resonance in contributing to the observed vibration behaviour at NDE motors unit 5 and 6.

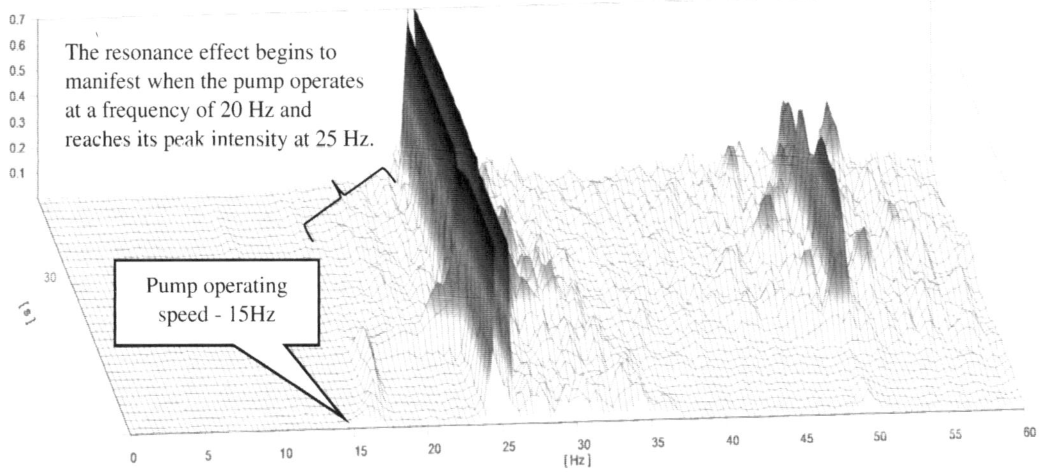

Figure 5: Run-up Waterfall Plot for NDE Motor unit 5

In summary, the investigation unequivocally identified the root cause of the "unacceptable" vibration observed at the motor NDE to be directly associated with the “normal” unbalance force generated by the motor's operating speed. This vibration was subsequently amplified due to the occurrence of resonance. Specifically, resonance transpired when the natural frequency of the motor NDE (25 Hz) coincided precisely with the motor's operating speed. The presence of resonance at the NDE was further confirmed through the comprehensive analysis of the waterfall plot derived from the run-up tests. This graphical representation effectively demonstrated the occurrence of resonance as the motor speed progressively approached its steady-state operating speed. These findings provide a compelling explanation for the excessive vibration observed at the NDE of the motors and validate the role of resonance in amplifying the vibration levels.

REMEDIAL ACTIONS IMPLEMENTED:

The most effective approach for vibration control was therefore to eliminate the resonance by shifting the natural frequency of the motor NDE away from the problematic 25 Hz frequency. To achieve this, a viable solution entailed reinforcing the motor pedestal to increase its stiffness. Consequently, it was proposed to reinforce the motor pedestal by adding 30mm thick ribs at both openings of the existing pedestal, as depicted in Figure 6. The reinforcement design aimed to shift the primary natural frequency of the motor-pedestal system from the original 25 Hz to 42 Hz, as determined through Finite Element Analysis (FEA) results. Photographs of the actual installation of the additional ribs at the pedestal is shown in Figure 7.

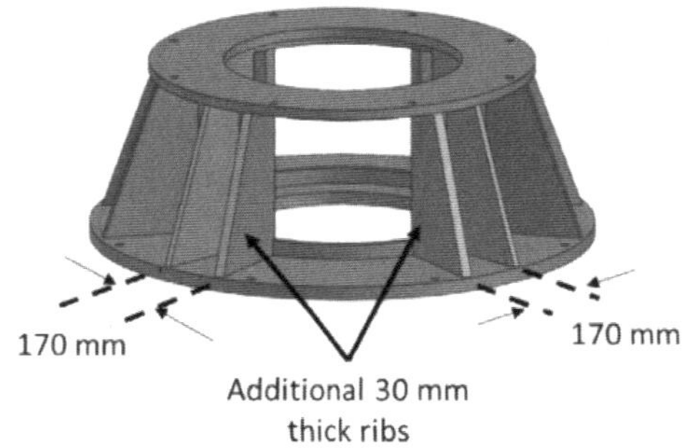

Reinforcement of the motor pedestal with additional 30mm thick ribs at both opening of the existing pedestal

Figure 6: Design of the Reinforcement for the motor pedestal

a) Original Pedestal of Motor (before modification)

b) Pedestal after the addition of new reinforce ribs

Figure 7: Reinforcement of the motor pedestal with additional 30mm thick ribs at both opening of the existing pedestal

Despite the fabrication, installation, and testing of the new pedestal, this particular attempt unfortunately did not succeed in reducing the excessive motor vibration as expected. The method employed did not result in a significant shift in the natural frequency of the motor-pedestal-floor system. Consequently, the vibration amplitude of Pump unit 6 remained persistently high even after the modification of the motor's pedestal. The outcome highlights the limitations of the chosen approach and suggests that alternative measures need to be explored in order to effectively address the excessive vibration issue.

The subsequent remedial measure proposed was the installation of a tuned mass damper (TMD) to effectively absorb the resonance-induced vibrations. The utilization of TMDs for vibration control in structures and machinery has been extensively documented in various literature sources, including references [1] to [4].

In this study, Finite Element Analysis (FEA) was employed to design a custom-made TMD specifically tailored for attachment at the motor NDE. The FEA allowed for precise tuning and modifications to optimize the TMD design. After several iterations of fine-tuning and adjustments, the final configuration of the TMD was successfully installed at the motor NDE, as illustrated in Figure 8.

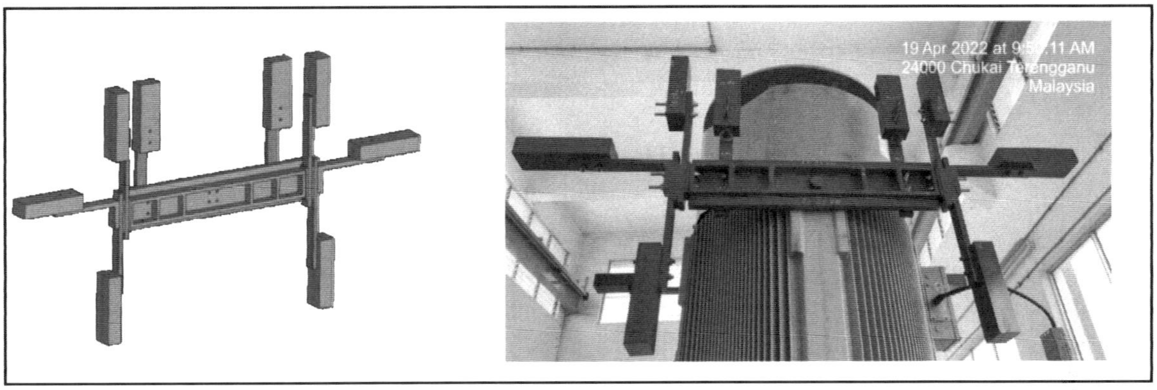

Figure 8: TMD installed at the motor NDE

As a result, the vibration levels for motor unit 6 significantly decreased from 18 mm/s to less than 5 mm/s, while for motor unit 5, the vibration levels dropped from 12 mm/s to less than 5 mm/s. These notable reductions demonstrate the efficacy of the implemented TMD solution in mitigating the excessive vibrations observed at the NDE of these motors. The application of the proposed TMD solution ensured that the vibration levels at the motor NDE complied with the ISO 10816-3 standard.

In short, The TMD serves as a means to mitigate the vibration levels and effectively address the resonance issue that the previous approach was unable to resolve.

CONCLUSION:

The study demonstrates that the installation of a properly designed tuned mass damper (TMD) serves as an effective means of dissipating vibration energy associated with resonance. While the TMD does not completely eliminate the occurrence of resonance in the machinery, it significantly reduces the severity of vibration levels experienced during resonance conditions. This case study reinforces the notion that TMDs can provide a valuable and accessible tool in vibration control strategies, allowing for enhanced performance and reduced potential for machinery failures due to resonance-induced vibrations. The paper serves as a reference for industry professionals seeking effective and feasible measures for vibration control in scenarios where resonance avoidance becomes challenging.

ACKNOWLEDGMENTS:

The authors express their heartfelt gratitude to the Institute of Noise and Vibration at the Universiti Teknologi Malaysia (UTM) for providing funding support for this study through the Higher Institution Centre of Excellence (HICoE) Grant Scheme (R.K130000.7809.4J226, R.K130000.7843.4J227, and R.K130000.7843.4J228).

REFERENCES:

[1] Liangkun Wang, Weixing Shi, Quanwu Zhang, Ying Zhou, "Study on adaptive-passive multiple tuned mass damper with variable mass for a large-span floor structure", Engineering Structures, Volume 209, 2020, 110010, ISSN 0141-0296.

[2] Yiqing Yang, Wei Dai, Qiang Liu, "Design and implementation of two-degree-of-freedom tuned mass damper in milling vibration mitigation", Journal of Sound and Vibration, Volume 335, 2015, Pages 78-88, ISSN 0022-460X.

[3] Said Elias, Vasant Matsagar, "Research developments in vibration control of structures using passive tuned mass dampers", Annual Reviews in Control, Volume 44, 2017, Pages 129-156, ISSN 1367-5788.

[4] Yung-Hsiang Chen, Yen-Hui Huang, "Timoshenko beam with tuned mass dampers and its design curves", Journal of Sound and Vibration, Volume 278, Issues 4–5, 2004, Pages 873-888, ISSN 0022-460X.

ANNUAL CONFERENCE

Where Theory meets Practice

EXPERIMENTAL STUDY ON DEBONDING DEFECT DETECTION IN A LARGE-SCALE CONCRETE-FILLED STEEL TUBE MEMBER BASED ON SURFACE STRESS WAVE MEASUREMENTS

Jiang Wang, PhD candidate
College of Civil Engineering, Huaqiao University
Jimei Avenue 668, Xiamen 361021, China

Main research fields: Study on nondestructive testing method and mechanism of concrete-filled steel tube by using guided wave technique.

Qian Liu, PhD candidate
College of Civil Engineering, Huaqiao University
Jimei Avenue 668, Xiamen 361021, China

Main research fields: Study on nondestructive testing method and mechanism of concrete-filled steel tube by using impedance technique.

Bin Xu, PhD, Professor and Director
College of Civil Engineering, Huaqiao University
Key Laboratory for Structural Engineering and Disaster Prevention of Fujian Province (Huaqiao University)
Jimei Avenue 668, Xiamen 361021, China

Main research fields: Study on new technology and mechanism of nondestructive testing of civil engineering structures; Identification of structural nonlinear behavior; New wind support structure system; Multi-disaster protection; Multi-scale and multi-physics field simulation of concrete materials and components

ABSTRACT:
Recently, large-scale circular concrete-filled steel tube (CFST) members with a diameter of 3.6m have been used in a 350m Steel-box-CFST composite towers of a long-span suspension bridge in China with a main span of 2,300m. Interface debonding defect between steel tube and concrete core of CFST members possibly occurs after the hardening of mass concrete. It has been a critical need to detect the inaccessible interface debonding defect because it negatively affects the mechanical behavior of CFST members. In this paper, an active interface debonding detection method for a large-scale CFST members using surface stress wave measurement is firstly proposed, where Piezoelectric-Lead-Zirconate-Titanate (PZT) patches are mounted on the outer surface of the CFST specimen as both actuator and sensor. Number of debonding defects with different dimensions are mimicked in the specimen. Surface wave under both continuous sinusoidal and modulated pulse excitations is measured in each pitch-catch configuration. The abnormal voltage amplitude and wavelet packet energy of the surface wave measurement under continuous sinusoidal excitation and modulated pulse excitation are determined. Then, the mimicked interface debonding defects is determined. Test results show that the interface debonding defect in the CFST specimen can be detected successfully using surface wave measurements.

KEY WORDS: Concrete-filled steel tube(CFST); Interface debonding defect; Piezoelectric-Lead-Zirconate-Titanate; Surface wave measurement; Wavelet packet energy; Experimental study.

INTRODUCTION:
Concrete-filled steel tube (CFST) member is commonly used as the main vertical and axial load-bearing component in super high-rise buildings and long-span bridges. The confinement effect of steel tube on concrete core of CFST members is the key factor in improving its mechanical properties such as load-carrying capacity and ductility. However, interface debonding defects between concrete core and steel tube in large-section CFST occur easily due to unavoidable shrinkage of mass concrete, uneven distribution of hydration heat and poor concrete construction technology. Interface debonding defects directly weaken the confinement effect of steel tube on concrete core, and thus produce negative effects on the load-carrying capacity and ductility of CFST members [1, 2]. Therefore, it is particularly critical to develop effective interface debonding defect detection methods for large-section CFST members.

For different defects such as cracks and debonding between concrete and reinforcement in reinforced concrete (RC) structures, nondestructive testing (NDT) technologies have been proposed [3, 4], including electromagnetic method [5], infrared thermal imaging [6], ultrasonic method and X-ray method [7]. However, due to the electromagnetic shielding effect of steel tubes and the complexity of internal structure in CFST members, most of the existing NDT methods cannot effectively detect the interface debonding defects in CFSTs. Literature [8] states that traditional ultrasonic method is only applicable for defect detection in concrete core of CFSTs when the steel tube is well bonded to concrete, and the detection method of interface debonding defect in CFSTs has not been proposed. Since stress wave propagation in CFSTs is not affected by electromagnetic shielding effect of steel tubes [9-12], the defects detection method with stress wave measurement in CFST members has attracted wide attention recently.

To detect interface debonding defects in CFSTs, Xu et al. [12] firstly proposed an active detection method with PZT-based stress wave measurement using actuating and sensing functions of PZT material, and verified the effectiveness of this method through experiments and numerical simulation. To study the mechanism of structure defect detection methods using stress wave measurement with PZT materials, multi-physics simulations on stress wave propagation and PZT sensor response have been carried out considering the coupling effect between PZT material and CFST structure and the positive and inverse piezoelectric effects of PZT material [13-17]. Xu et al. [18, 19] established a multi-physics coupling finite element model (FEM) of rectangular and circular CFST columns and PZT materials, and studied the propagation of stress waves in CFST members through transient dynamic analysis. The influences of the frequency and amplitude of excitation signals, the distance between the PZT actuator and the embedded PZT sensor and the size of defect on the stress wave propagation and the output signal of embedded PZT sensors are discussed. Unfortunately, the installation of embedded PZT sensors in CFST members is inconvenient and is not suitable for the defect detection of CFST members in existing buildings. Chen et al. [20] used mesoscale numerical transient analysis to study the influence of location and length of interface debonding defects in CFST on the propagation of multi-channel surface waves, and verified the feasibility of interface debonding detection using multi-channel surface waves. In order to further verify the effectiveness of the surface wave measurement method, Chen et al. [21] carried out multi-physical field coupling numerical simulations to analyze the stress wave propagation mechanism in CFSTs with different interface debonding defects, and the simulation results showed that surface wave measurement could effectively detect the interface debonding defects in CFST. An experimental study on the feasibility of interface debonding defects detection approach with surface wave measurement for large-scale CFST members is desired.

To further study the effect of interface debonding defects at different locations and with different sizes in large-section CFST members on surface wave measurement and to investigate the feasibility of interface debonding detection approach using PZT-based surface wave measurement, in this paper, experimental study on the interface debonding detection with a full-scale CFST column with number of artificially mimicked interface debonding defects was carried out. Interface debonding defects with different sizes were mimicked at certain locations of the inner surface of the steel tube of the specimen. Then, number PZT patches were pasted on the outer surface of steel tube as actuators and sensors. The response of each PZT sensor was measured using a pitch-catch configuration under both continuous sinusoidal and modulated pulse excitations. The amplitude and wavelet packet energy of each PZT response of different measurement paths were analyzed, and the influence of interface debonding defects on the amplitude and wavelet packet energy of PZT sensor measurements was obtained. The results show that the interface debonding obviously influences surface wave measurement, and the surface wave measurement can be used to effectively detect the interface debonding defects of large-scale CFST columns.

PRINCIPLE OF INTERFACE DEBONDING DETECTION WITH SURFACE WAVE:

The principle of interface debonding defect detection for a CFST column using surface wave measurement with surface-mounted PZT actuating and sensing technologies is shown in Fig. 1.

A pair of PZT patches are pasted on the outer surface of the steel tube of the CFST specimen at a certain distance and are used as an actuator or a sensor respectively. First, the PZT actuator is excited by a high-frequency electrical signal to generate mechanical vibration on the outer surface of the steel tube due to the inverse piezoelectric effect, thus generating stress waves propagating along the steel tube in CFST specimen. The surface wave propagating along the steel tube is measured by the corresponding PZT sensor along each pitch-catch measurement configuration. In each pitch-catch measuring path between the PZT actuator and sensor, when the stress wave propagating along the steel tube encounters an interface debonding defect, the energy transfer from the steel tube to the concrete core decreases, and the energy consumption of stress wave also decreases. When the stress wave continues to propagate along the steel tube to PZT sensor, PZT sensor converts the vibration into a voltage signal due to the positive piezoelectric effect, and its response will be greater than that of the PZT sensor on the measurement path without an interface debonding defect. Therefore, the interface debonding defects can be detected by comparing the output signals of PZT sensors along the healthy measurement path and that of the debonding measurement path in the CFST structure.

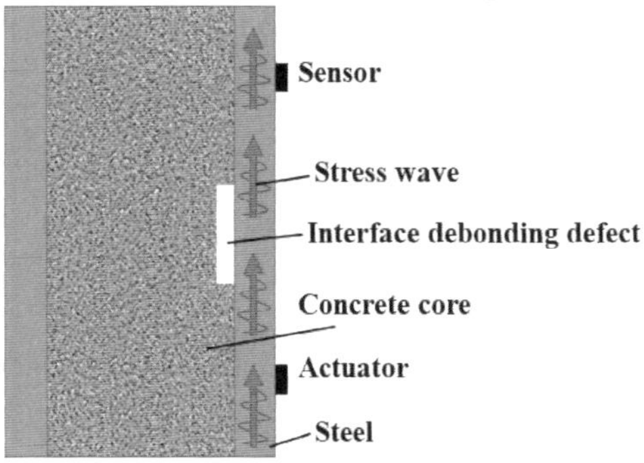

Fig 1. Debonding detection principle of CFST using PZT actuator and sensor

DEFECT DETECTION TEST FOR FULL-SIZE CFST SPECIMEN:

In this test, a full-size circular cross-section CFST column specimen with a height of 600mm and an outer diameter of 3,600mm was designed and constructed. The thickness of steel tube is 20mm. The type of steel tube is Q235 and high-strength self-compacting concrete with a strength grade of C60 is poured. To consider the effect of the interface debonding defects under the inner horizontal diaphragm on surface wave propagation, several artificial mimicked interface debonding defects with different sizes were designed in the specimen.

As shown in Fig. 2, three cross-sections A, B and C are set in the height direction of CFST column, where B is the cross-section where the center point of artificial interface debonding defect is located. The number and dimension of each artificially mimicked interface debonding defect are shown in Table 1. For the convenience of expression, the expanded view of the CFST specimen was divided into 16 equal parts. The specific location of each defect is shown in Fig. 2 (a). PZT actuators and sensors are located at cross-sections A and C . Sixteen PZT patches are arranged along the circumferential direction of cross-sections A and C, respectively. The numbering method of each PZT is unified by adding numbers 1-16 after the name of each cross-section, as shown in Fig.2 (b) - (c).

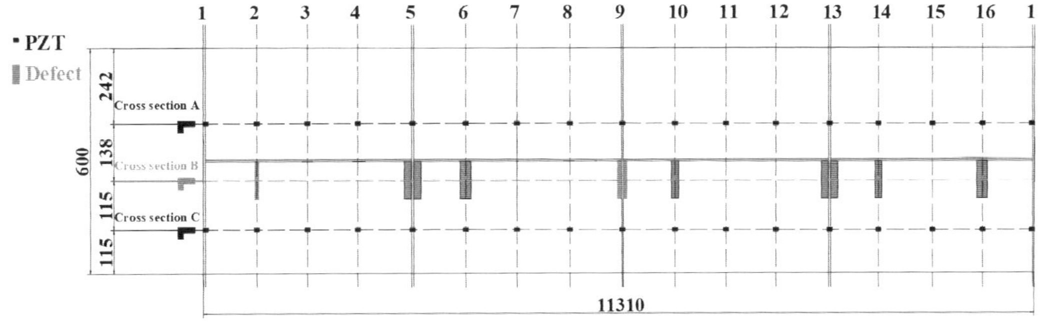

(a) Expanded view of CFST and locations of PZT patches and simulated defect

(b) PZT arrangement at cross-section A (c) PZT arrangement at cross-section C

Fig. 2 Defects and PZT arrangement in full-size CFST specimen

Table 1 Mimicked debonding defects and their dimension

	Number	BD1	BD2	BD3	BD4	BD5	BD6	BD7	BD8
Directions	Radial (mm)	3	3	3	5	5	5	7	7
	Circumferential (mm)	50	100	150	50	100	150	100	150
	Vertical (mm)	100	100	100	100	100	100	100	100

To realize the interface defect detection on cross-section B, the PZT patches on cross-section A are used as actuators, and the PZT patches on the same vertical line on cross-section C are used as sensors to measure surface waves using the pitch-catch configuration. The actual interface debonding defects at the steel tube are shown in Fig. 3, where each interface debonding defect is mimicked by pasting a soft ethylene vinyl acetate (EVA) foam plate at a predetermined position of the inner surface of the steel tube with epoxy resin.

The surface wave measuring system includes a signal generator, high-speed data acquisition instrument and notebook workstation. Fig. 4 shows the measurement system and the CFST specimen installed with PZT actuators and sensors. In the test study, a continuous sinusoidal excitation signal with a frequency of 200kHz and a five-cycle pulse excitation signal with a center frequency of 100kHz shown in Eq. (1) are used as input signals to excite PZT actuators located on cross-section C.

$$\begin{cases} V=V_0 \sin(2\pi ft)\,(\sin(2\pi ft/10))^2 & t \geqq 5T \\ V=0 & t \leqq 5T \end{cases} \tag{1}$$

where V is the voltage amplitude of excitation signal, V_0 is the voltage amplitude of excitation signal, f is the frequency, t is the time step, and T is the reciprocal of frequency.

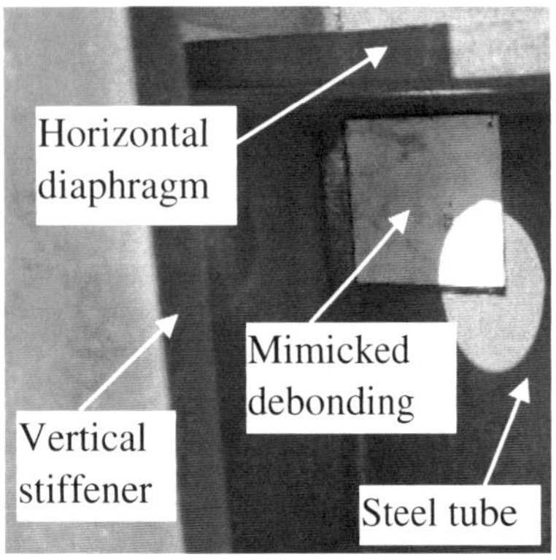

Fig. 3 Actual defect of BD 2

Fig. 4 Test system

TEST RESULTS ANALYSIS:
The steady-state amplitudes and the wavelet packet energy values of the sensor measurements under the continuous sinusoidal excitation signal and the five-period pulse signal, respectively, in different measuring paths are compared to evaluate the interface bonding condition. The measurement results of PZT sensors on all 16 test paths are shown in Fig. 5. According to the statistics-based anomaly data analysis and abnormal values judgment method described in literature [8], the abnormal threshold of amplitude and wavelet packet energy values of the measured signals of PZT sensors are determined.

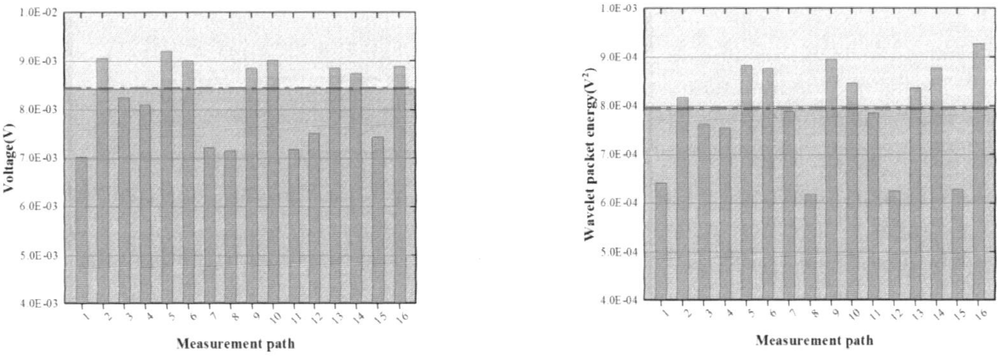

(a) Steady-state amplitude under sinusoidal excitation

(b) Wavelet packet energy under pulse excitation

Fig. 5 Measurement result and abnormal values judgement

According to the above data analysis method, it is determined that the abnormal threshold of steady-state voltage amplitude under continuous sinusoidal excitation is 0.0085V, which is represented by the red horizontal dotted line in Fig. 5 (a). On the measurement path of interface debonding defects BD1-BD8, the steady-state voltage amplitudes of output signals collected by PZT sensors are all located in the green background area and are higher than the threshold shown as a red horizontal line. Results show that all mimicked interface debonding defects under the horizontal diaphragm can be accurately identified by surface wave measurement. The wavelet packet energy value of output signal of PZT sensor of each measurement path under a five-cycle pulse excitation is compared and analyzed, and the abnormal threshold of all wavelet packet energy values is obtained, which is $0.0008V^2$, as shown in the red horizontal dotted line in Fig. 5 (b). Results show the wavelet packet energy value of output signals measured by PZT sensors on the measurement paths with an interface debonding defect between cross sections A and C is significantly higher than the abnormal threshold.

Under the continuous sinusoidal excitation signal and the five-period pulse excitation signal, the interfacial debonding defects below the horizontal diaphragm in tower CFST specimen can be accurately detected using surface wave measurement.

Relation between wavelet packet energy and debonding area and thickness:
To discuss the possibility of identify the debonding defect level using the surface wave measurement, further investigation on the quantitative relationship between interface debonding area and the depth and the wavelet packet energy values of PZT sensors response at different mimicked interface debonding defects is carried out. The relationship between the interface debonding defect area (product of the dimensions in circumferential and vertical directions) and defect radial thickness and the corresponding wavelet packet energies is fitted into a three-dimensional surface shown in Fig. 6.

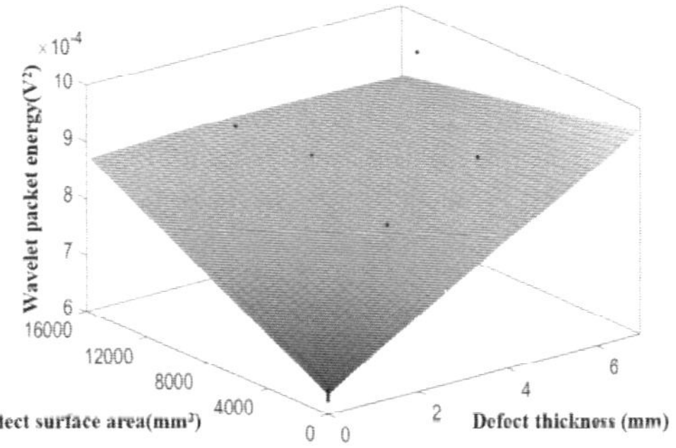

Fig. 6 Wavelet packet energy corresponding to defects of different sizes

The relationship between the wavelet packet energy value of the PZT sensor measurement signal and the defect size is shown in Eq. (2).

$$y=6.348e{-}4-5.043e{-}5x_1+1.516e{-}8x_2-5.749e{-}7x_1^2-2.918e{-}9x_1x_2 \quad (2)$$

where x_1 is the radial thickness of defect, x_2 is the surface area of defect, that is, the product of the circumferential height and vertical height, and y is the wavelet packet energy value of output signal measured by PZT sensor. The correlation coefficient used to judge the degree of correlation between the measured data is 0.9501, and the correlation coefficient is closer to 1. As can be seen from Fig. 6, the wavelet packet energy is related to the size of defect. The corresponding wavelet packet energy also increases with the increase of defect size.

CONCLUSION:

In this paper, the effectiveness of an interface debonding defects detection method using surface stress wave measurement with PZT actuating and sensing technologies is verified by experiments. The stress wave was measured by PZT sensors pasted on the outer surface of the steel tube of the CFST specimen and PZT actuators are also on the outer surface of CFST specimen. Under the continuous sinusoidal signal and five-period pulse excitation signal, the amplitude and wavelet packet energy values of the surface wave measured by PZT sensors of different wave travelling paths were analyzed. The mimicked interface debonding defects are detected successfully. Based on the experimental results of this paper, the following conclusions can be made.

(1) Under either the continuous sinusoidal excitation signal or the five-period pulse excitation signal, the amplitudes and wavelet packet energy values of the surface wave measurements by PZT sensors output signals on all measurement paths with interface debonding defect are significantly different from them of the PZT sensor corresponding to the measurement paths without interface debonding defect.

(2) Using statistics-based anomaly data analysis approach, the abnormal thresholds of the amplitude and wavelet packet energy value of surface wave measurement under two kinds of excitation signals are obtained. Using the amplitudes and wavelet packet energy values of the surface wave measurements from PZT sensors can be used to effectively judge the location of the interface debonding defect. Moreover, the proposed debonding detection approach using surface wave measurement is sensitive to interface debonding defects with small thicknesses in the radial direction of the CFST specimen.

(3) Based on the experimental results, the relationship between the wavelet packet energy value and the interface debonding area and thickness is determined, which can be potentially used to estimate the level of debonding for CFST members in practice.

REFERENCES:

[1] Xu, B., Zhang, T., Song, G.B., et al. "Active interface debond detection of a concrete-filled steel tube with piezoelectric technologies using wavelet packet analysis", Mechanical Systems and Signal Processing, 2013, 36 (1): 7-17
[2] Chen, H.B., Nie, X., Gan, S.Y., et al. "Interfacial imperfection detection for steel-concrete composite structures using NDT techniques: A state-of-the-art review", Engineering Structures, 2021, 245: 112778.
[3] Chen, H.B., Zhou, M., Gan, S.Y., et al. "Review of wave method-based non-destructive testing for steel-concrete composite structures: Multi-scale simulation and multi-physics coupling analysis", Construction and Building Materials, 2021, 302: 123832.
[4] Liu, Z., Chen, X., Wu, P., et al. "Investigation on micro-structure of self-compacting concrete modified by recycled grinded tire rubber based on X-ray computed tomography technology", Journal of Cleaner Production, 2021, 290: 125838.
[5] Sun, Y.H., Liu, S.W., Deng, Z.Y., et al. "Magnetic flux leakage structural health monitoring of concrete rebar using an open electromagnetic excitation technique", Structural Health Monitoring, 2017, 17 (2): 121-134.
[6] Hiasa, S., Catbas, F.N., Matsumoto, M., et al. "Monitoring concrete bridge decks using infrared thermography with high speed vehicles", Structural Monitoring and Maintenance, 2016, 3 (3): 277-296.

[7] Chen, J., Yang, C.L., Wang, Q.D., et al. "Nonlinear ultrasonic second harmonic assessment of concrete defects based on embedded piezoelectric sensors", Research in Nondestructive Evaluation, 2020, 31 (4): 354-270.
[8] Technical Specification for inspection of concrete defects by ultrasonic method, CECS 21:2000
[9] Chen, Q., Xu, B., Chen, H.B., "Simulation analysis of detection of interface debonding in concrete-filled steel tube based on surface wave", Piezoelectrics and Acoustoopics, 2019, 41(1): 80-84.
[10] Lu, L., Yu, H., Qiao, P., "Nondestructive evaluation of concrete based on stress wave propagation mechanism: A Review", Chinese Quarterly of Mechanica, 2021, 42(2): 197-216.
[11] Fan, X., Xu, B., Xia, S., "Number simulation study on stress wave propagation in grouting sleeve with defects", Piezoelectrics and Acoustoopics, 2019, 41(6):793-796.
[12] Qiao, L.X., Xiao, Y., Luo, X.S., et al. "Experiment on defect detection for grouting splice sleeves based on piezoelectric wave measurement method", Piezoelectrics and Acoustoopics, 2022.
[13] Xu, B., Li, B., Song, G., "Active debonding detection for large rectangular CFSTs based on wavelet packet energy spectrum with piezoceramics", Journal of Structural Engineering, 2012, 139 : 1435-1443.
[14] Mo, X.P., "Simulation and analysis of acoustics transducers using the ANSYS software", Technical Acoustics, 2007, 26(6):1279-1290.
[15] Xu, B., Chen, H.B., Mo, Y.L., et al. "Multi-physical field guided wave simulation for circular concrete-filled steel tubes coupled with piezoelectric patches considering debond defects", International Journal of Solids and Structures, 2017, 122: 25-32.
[16] Xu, B., Chen, H.B., Mo, Y.L., et al. "Dominance of debond defect of CFST on PZT sensor response considering the meso-scale structure of concrete with multi-scale simulation", Mechanical Systems and Signal Processing, 2018, 107: 515-528.
[17] Wang, J., Xu, B., Chen, H.B., et al. "Mesoscale numerical analysis and test on the effect of debond defect of rectangular CFSTs on wave propagation with a homogenization method", Mechanical Systems and Signal Processing, 2022, 163: 108135.
[18] Wang, J., Xu, B., Chen, H.B., et al. "Multi-physics mesoscale substructure analysis on stress wave measurement within CFST-PZT coupling models for interface debond detection", Sensors, 2022, 22(3): 1039.
[19] Xu, B., Chen, H.B., Xia, S., "Numerical study on the mechanism of active interfacial debonding detection for rectangular CFSTs based on wavelet packet analysis with piezoceramics", Mechanical Systems and Signal Processing, 2017, 86: 108-121.
[20] Xu, B., Chen, H.B., Xia, S., "Wave propagation simulation and its wavelet package analysis for debonding detection of circular CFST members", Smart Materials and Structures, 2017, 19 : 181-194.
[21] Chen, H.B., Xu, B., Wang, J., et al. "Interfacial debonding detection for rectangular CFST using MASW method and its physical mechanism analysis at meso-level", Sensors. 2019, 19 (12): 2778.
[22] Chen, H.B., Xu, B., Zhou, T.M., et al. "Debonding detection for rectangular CFST using surface wave measurement: Test and multi-physical fields numerical simulation", Mechanical Systems and Signal Processing, 2019, 117: 238-254.

AT-MACHINE AND REMOTE TESTING OPTIONS FOR GREASE LUBRICATED MACHINERY

Rich Wurzbach, CLS, MLE
MRG Labs
Tribology Research Center at Knowledge Park
410 Kings Mill Rd
York, PA 17401

Rich Wurzbach has over 30 years of experience in the development of Condition Based Maintenance programs. He has been employed at several nuclear power plants and the US National Institutes of Health. He has served on two research teams in Denmark studying lubrication and sampling practices for offshore wind turbine components. Mr. Wurzbach holds four patents on his designs for lubricant sampling and analysis devices. He has presented and published over 70 scientific papers. He holds a BA in Chemistry from Millersville University (PA).

ABSTRACT:

Oil lubricated machinery has long been the focus of fluid analysis for prognostics and condition monitoring. Challenges due to sampling, and limitation on movement and flow of grease have placed grease-lubricated equipment at a disadvantage, relegating these assets to other secondary testing approaches such as vibration analysis or ultrasonic testing to evaluate lubricant and machine condition. However, more recent technology developments that have been standardized by ASTM result in more widespread use of laboratory-based grease analysis. Additional options are available for at-machine monitoring that can create near real-time condition assessment of grease lubricated equipment possible. In many cases, these portable instruments can screen large fleets of assets for overall wear and grease condition and allow screening for secondary testing at remote laboratories to gain a full assessment of grease condition and machine wear. This presentation will focus on the limitations and challenges for grease-lubricated assets, and present tiered solutions for enhancing reliability and grease optimization. Case studies will be provided for military assets, power generation machinery, and robotic automation applications.

KEY WORDS: Lubrication; Sensors; Grease analysis; Asset management.

INTRODUCTION:

Grease analysis technologies have undergone rapid development following the standardization of grease sampling with the American Society of Testing and Materials (ASTM) D7718 in 2011. This includes the development of new technology for testing greases directly (such as those instruments outlined in ASTM D7918) and the adaptation of existing oil analysis instruments to allow meaningful analysis of greases. Still, challenges exist in the handling of greases due to the very different flow properties that prevent typical pumping and suction methods used for oil samples. Additionally, greases present fundamental challenges to the development of in-situ sensor for machine deployment, because of the tendency of greases to adhere to surfaces, including sensor detecting surfaces. This paper will outline the important considerations for sampling greases from various machine applications, options for at-machine deployment of field instruments and sensors, and the strategies to leverage these technologies to optimize grease lubricated machine inspections and grease testing.

SAMPLING:

All measurements must begin with a careful consideration of the placement of measuring devices in order to produce measured quantities that have significance when applied to the target of interest. Some materials must be sampled from a location or device, and introduced to measuring sensor or processes to achieve this. Included in this latter group is heterogenous materials such as mineral deposits, certain fuels, and machine lubricants. In the case of lubricating oils, this sample challenge is generally reduced to limiting samples to those taken under a condition of turbulence, at a specific location within the machine that ensures uniformity of the sample and representative composition (i.e., upstream of filters or setting tanks, before dilution with lower concentration volumes of oil in the system).

In the case of greases, complexity of sampling heterogenous materials is ubiquitous, and made even more challenging by the non-Newtonian behavior of these fluids. Samples taken without considering these limitations often result in grease samples which do not represent the wear, contaminants, grease consistency or oxidative health. Many asset managers will use traditional oil sampling tools and methods (such as syringes or suction applied to immediately accessible grease in a machine). At this point, regardless of the analysis methods applied, the results will not add value to the understanding of existing machine and lubrication conditions.

ASTM Standard Practice for Obtaining In-Service Samples of Lubricating Grease (D7718) provides clear guidance to assess a sampling challenge, and determining the tools and methods required to ensure a small-volume but representative sample is obtained. Alternative approaches that generate larger volumes (sometimes required by instruments or labs to provide analysis) can results in composite samples that are actually a combination of the representative analytes, along with additional new or aged grease needed to meet the volume requirements, but diluting the accuracy of the sample.

AT-MACHINE TESTING:

While the design challenges of suitable in-situ sensors that can work with the unique flow properties of grease prevent current widespread adoption, for reasons already mentioned, there has been development of small footprint, portable, battery-operated devices that can be positioned at or near machinery for rapid determination of grease and machine condition. These devices can also include wireless data transfer, such as Bluetooth, enabling the handheld processing devices such as smartphones. These designs generally allow some level of analysis without requiring the use of chemicals, special power supplies, and other requirements found in laboratory instrumentation. Generally these devices produce simple numerical outputs that can be used with established criteria to screen samples and determine whether further, more detailed analysis is suggested.

Some of these devices are designed around sampling devices described in ASTM D7718 sampling standard, thus allowing a simple transition from the sampling to analysis steps, minimizing handling and the likelihood of introduction of handling errors. Examples exist both for machine monitoring tools and grease condition monitoring.

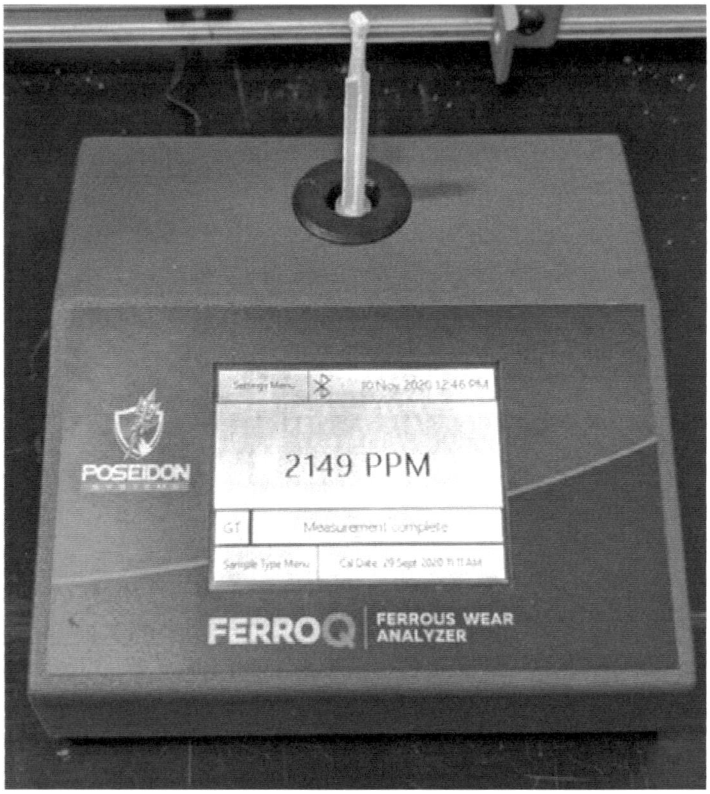

Figure 1: Grease Ferrous measurement

Figure 1 shows a ferrous detection device that works with the samples obtained in the Grease Thief device per ASTM D7718, and utilizes an inductive coil design to measure the total ferrous content. This design is an enhancement over previous oil ferrous sensor devices that depend on the sample being a homogeneous distribution of that ferrous wear, or requiring significant and time consuming (and not always effective) homogenization steps. This is accomplished by a signal processing method that measures the transit of the ferrous content as it enters the coil measuring area, and compensates for non-uniform particulate distribution. This is important as grease samples commonly contain distributions of various sized ferrous wear in a heterogeneous spatial distribution. Additionally, the measured results must be normalized for incomplete sample device fill, which is common due to the challenges of getting the semi-solid grease material into the sampling device. This normalization can be accomplished by weighing the sample and mass-normalizing the results, or using an effective length measurement of the Grease Thief sample based on the position of the piston that compresses the sample to the distal end of the sampling chamber.

Figure 2: Grease Colorimeter

Figure 2 shows a colorimeter device used to compare the grease optical properties to the new reference grease. This uses a Beer-Lambert Law relationship between spectral absorbance, optical pathlength, and analyte concentration. The Colorimeter uses precision spacers to allow compression between a reflective Teflon disc, and the spectrometer flat window. By creating a series of set pathlengths, greases with low, medium and high optical transmittance can be analyzed relative to the new greases, providing a numerical representation of the color spectrum utilizing CIE values. Comparison to the new grease reference produces a comparison CIE ΔE value, which can be compared to acceptance criteria developed for machine, application, grease type or environmental considerations.

Other such At-Machine sensors are under development, including Die Extrusion for grease consistency changes. Such devices used together as a Mini-Lab for grease screening can allow near immediate feedback for integration with other diagnostic inputs (vibration, ultrasound, infrared, process variables, etc.). This tiered approach to testing greases can expedite machine action decisions, prompting efficient use of secondary analysis capabilities, such as those available by shipping these samples to a laboratory.

GREASE LABORATORY ANALYSIS:

The ASTM Standard Test Method for evaluating in-service greases is D7918. This method uses the tests identified in TABLE 1 which list the key parameters measured and evaluated in the grease. Two additional tests (TABLE 2) are typically included in the analysis although they are not found in the current version of the ASTM D7918 standard. The Rotating Disc Electrode (RDE) Spectroscopy test and Fourier-Transform Infrared (FTIR) Spectroscopy test, provide elemental and molecular spectroscopy results, respectively, and complement the analysis performed in D7918 by monitoring non-ferrous wear particles, additives, contaminants, and the oxidation of the grease. While the ASTM D7918 method does provide for a particle counting test, a thin-film direct imaging technique, initially dark, greases with high solids levels, or very aged greases may prevent light transmission. The particle counting function can be substituted by monitoring the elemental Silicon (Si) or other element levels in the grease as measured by the Rotating Disc Electrode Spectrophotometer. For example, when monitoring helicopters, the most common source of external particulate is expected to be sand or dirt based on operating environments and action of air turbulence, Silicon is often a suitable indicative parameter to measure the ingression of these particles.

TABLE 1: ASTM D7918 Grease Analysis Tests

Grease Test	Parameters Measured	Value of Analysis
Ferrous Density	Parts per million (ppm) iron debris/wear; inductive coil method	Used to evaluate total accumulated wear between grease replenishments
Colorimetry	A spectral comparison to the new grease in the visible light range	Screen for mixing different greases, aging, and contaminants
Die Extrusion	A variable-rate shear flow of the grease at controlled temperature that compares the load profile to the new grease	Measurement of consistency changes that can predict tendency of grease to harden or soften in service and leak from housings
Linear Sweep Voltammetry	The "RULER" test quantifies the anti-oxidants found in the new grease, and compares the remaining levels and the time in service for the grease sample	Can evaluate the remaining oxidative life of the grease to confirm ability to continue for longer service intervals without degradation
Moisture	Parts per million water is measured by thermal transfer from a grease thin-film and measurement with a humidity sensor	Replenishment of grease can flush contaminants; this can ensure excessive moisture will not intrude at extended service intervals
Particle Contaminants	Direct imaging particle count method to quantify in mg/g the amount of particles >10, >25 and >80 microns. Monitoring silicon (Si), aluminum (Al) and other key elements characteristic of soils or ambient contaminants as well.	Replenishment of grease can flush contaminants; this can ensure excessive particulate will not intrude at extended service intervals

A streamlined version of the grease analysis process is sometimes used to provide two key analysis parameters to screen large numbers of samples for outliers. The approximate 1-gram sample size taken in the Grease Thief can be efficiently evaluated for wear content (ferrous debris in ppm) and the optical profile (colorimetric spectral response). The wear content can be trended and evaluated against criteria, and statistics used to determine action levels for maintenance and intervention. The colorimetric spectrum can be compared to the new, fresh grease to identify potential cases of grease mixing, aging and other contaminants. Together these two screening tests can efficiently and cost-effectively monitor a large number of assets to create statistical norms and outliner criteria, to determine those samples that should be designated for more detailed analysis.

TABLE 2: Additional Grease Tests performed, not in ASTM D7918

Grease Test	Parameters Measured	Value of Analysis
FTIR	ASTM E2412 is widely used in the US Military Joint Oil Analysis Program (JOAP) to detect contaminants and changing oil condition. This Fourier-transform Infrared Spectroscopy evaluates the organic functional groups in the sample.	For greases, this can be used to compare to the new grease formulation to find the presence of different greases mixing, contaminants, and oxidation of the grease. Will help to determine if oxidation is occurring in the existing or projected service intervals. May also be helpful to detect some anti-oxidant additives, depending on the specific grease product formulations, and differentiate thickener chemistries.
Rotating Disc Electrode (RDE) Spectroscopy	Parts per million measurement of elemental metals as wear, additives and contaminants, using a solvent-preparation method, and ASTM D6595 standard.	Will complement ferrous debris monitoring for wear measurement of non- ferrous components (such as copper), oxidized ferrous particles such as rust, corrosion and fretting wear, help determine relative sizes of ferrous wear by comparing elemental Fe to ferrous levels.

When the detailed analysis is performed, four key areas are evaluated among the up to 10 tests possible with the 1-gram sample. They are Wear, Consistency, Contamination, and Oxidation. Together they provide a clear picture of the condition of both the machine wear, and the lubricant itself. It creates the opportunity to move away from usage-based or calendar-based intervals for re-greasing, which rely on multiple assumptions about the machine design, temperatures that the grease experiences, vibration conditions (that lead to grease shear and separation), and contaminant types and amounts, that dictate the need to flush out old grease.

Sustainability initiatives drive organizations to review historical practices, and determine opportunities to reduce waste, consumption and pollution. The widespread use of excessive amounts of grease, driven by a "more is better" approach, or logic that its cheaper to buy grease than to replace machinery, creates a very inefficient use of resources. In each case where evaluations have been made to utilize grease analysis to drive an optimized approach to greasing activities, a reduction of 50% of grease consumed has been the norm, as shared in several case studies below. In addition to reduced grease consumption, greater insight is gained into the condition of grease lubricated components, complementing other diagnostic data such as vibration analysis, ultrasound and thermal imaging, with an added benefit of determining root-cause of lubrication failures. Overall, machine life is extended through optimal lubrication conditions, labor required to lubricate and maintain is reduced, and a more sustainable operational profile of the lubricated assets is achieved.

CASE STUDIES:

Locomotive Wheel Bearings

Studies were initiated by TTCI (Transportation Technology Center, Inc.), the research arm of the Federal Railroad Administration, to test grease samples from locomotive wheel bearings, and shortly thereafter adopted by locomotive manufacturers to supplement other diagnostic monitoring activities. The key focus was on obtaining a representative sample, which is difficult based on the accessibility of the wheel axle U-tubes. A tool and method was developed and tested and found to correlate highly to as-found conditions upon disassembly of the wheel bearings. The primary analysis in this case being the utilization of the ferrous debris tester shown in Figure 1, automated to utilize a six-axis robot for handling the sample, reading the barcode, and weighing the sample. This autonomous analysis solution includes mass-normalization calculations, and real-time upload of data to the cloud, even after hours and lights-out conditions in the lab.

Using this approach, over 30,000 grease samples have been analyzed and used to diagnose and route locomotives with degraded wheel bearings to maintenance depots for timely disassembly and inspection. Having several years history of these activities, it was calculated for one Class 1 railroad that the monitored locomotive "set-out" rate (on-track failures requiring removal and replacement of the locomotive) had been reduced by 50%, resulting an a monthly savings of $500,000 on a program cost of less than $35,000.

Helicopter

Boeing and MRG Labs conducted a Lubrication Optimization Study (LOS) on the H-47 Chinook helicopter to establish the engineering basis and artifacts to define the grease lubrication intervals for selected Drive, Rotor, and Landing Gear components. The study's main focus was a laboratory analysis of grease samples collected from these components, to characterize how wear, properties and contaminants change as time and aircraft hours accumulate. Historical basis for greasing intervals extended back to the early 1960's and the original design of the aircraft. In the ensuing years, some changes were made to greasing intervals, but mostly in response to failures or conditions observed during maintenance, which would consistently result in shortening the frequencies previously defined.

To ensure there was statistical significance in the analysis, almost 1200 samples were collected in two phases from May of 2018 through July of 2019. Five different operators participated in collecting samples. This level of participation provided an opportunity to quantify a diverse range of environmental operating conditions such as "hot", "dry", and/or "salty".
MRG Labs evaluated the grease samples for Ferrous Density (FdM), Colorimetry, Die Extrusion (shear flow), Linear Sweep Voltammetry (anti-oxidation properties), Moisture, Particle Contaminants (external sources such as Silicon, Aluminum), Fourier-transform Infrared Spectroscopy (organic contaminants), and Rotating Disc Electrode (RDE) Spectroscopy (wear measurement of non-ferrous components). Oxidation was the primary limiting parameter in most cases, while the secondary limiting parameter were more dispersed.

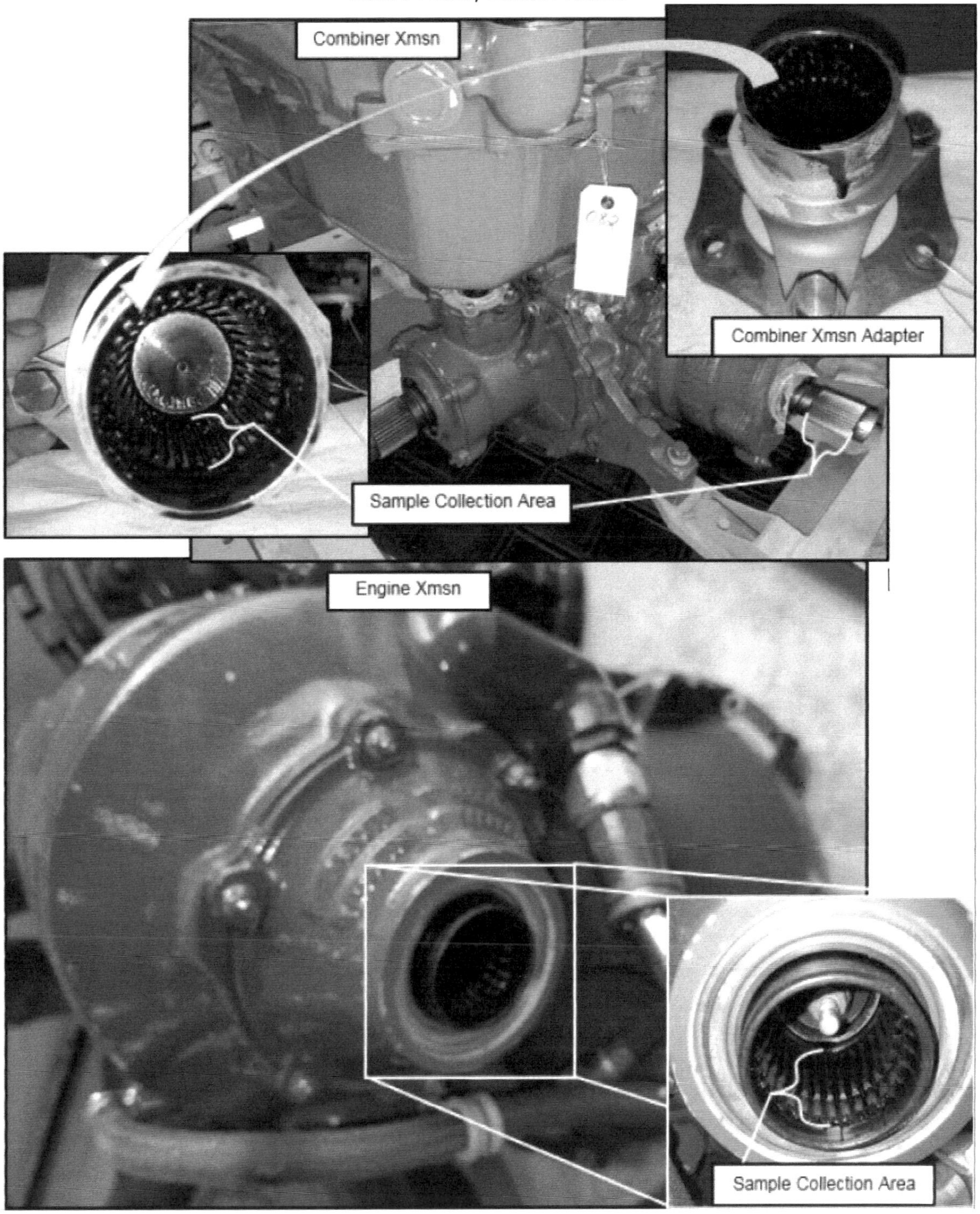

Figure 3: Sampling Diagram for Cross Shaft Engine and Combiner Transmission Adapters

Boeing conducted a field data analysis on each of the sampled components. This analysis supplemented the grease analysis conducted by MRG Labs. Together these analyses produced the engineering basis for the recommended servicing interval for the selected Drive, Rotor, and Landing Gear components. An example of the use of data and extrapolation of the end of useful service life of the grease is shown in Figure 5.

As a result of the study the majority of the lubrication intervals were extended, many double the established grease service intervals currently fielded by the various operators. This extension provides each operator with an opportunity to adjust their grease servicing intervals thus allowing a reduction in down time and maintenance costs while ensuring the continued safe operation of the aircraft. The benefit of the study is that final service interval determinations can be aligned with existing maintenance intervals, and to support a maintenance optimization effort completed by Boeing.
These results translate to fewer interruptions to operations for required servicing. In total, an estimated reduction from 98 lubrication tasks per 1000 flight hours to just 56 lubrication tasks per 1000 flight hours, and a reduction from 20 service interruptions per 1000 flight hours to just 10 service interruptions per 1000 flight hours. Across the entire Chinook fleet, adoption of these optimized greasing intervals by all operators will result in an annual savings of approximately US$100 million per year.

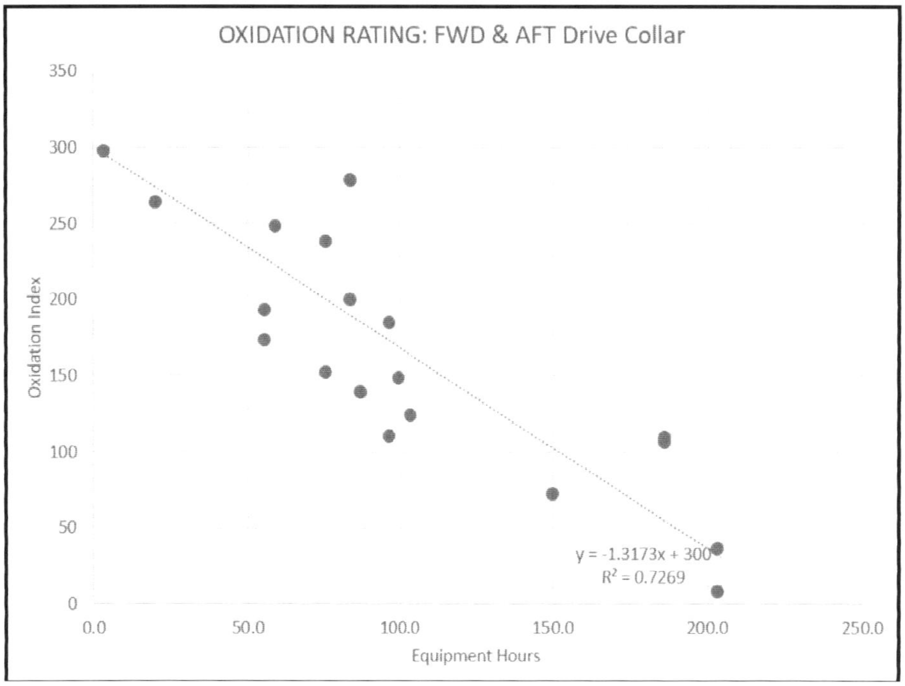

FIGURE 4: Drive Collar samples Oxidation Rating

Robots

Six-axis assembly robots are widely used in auto manufacturing. The use of robotics and automation continues to grow more prevalent across many industries, as companies are realizing the full potential of its ability to reduce costs, boost production, and deliver greater efficiency and competitiveness. Typical industrial robotic applications include welding, painting, assembly, product inspection, and testing. Bearing life and performance demand that the correct type, material, and lubrication are specified. The advantages of robots are precision, speed, and strength. These demands challenge the bearings and, ultimately, their lubricants.

Most robot operating manuals dictate that grease must be flushed and replaced annually or more often, but this may not always be necessary. It is important to give consideration to the operating cycles, local temperatures (some robots are near furnaces, others in air-conditioned spaces), load, moment arm and acceleration (can cause wildly varying stresses on the gears and grease), and the ambient environment (dust, paint, chemicals, etc.) If Preventive Maintenance (PM) task frequency does not reflect changing variables, then it will inevitably be either overly conservative or ineffective. By adopting a Condition Based Maintenance (CBM) approach using grease sampling and analysis, the maintenance intervals can be optimized, and the maintenance resources for reliable robot operation can be used efficiently.

For a major US auto manufacturer, a fleet of robot performing varying tasks was targeted for a grease analysis study. OEM guidance for grease replenishment was uniform, despite the same robots being deployed in widely varying applications, including large movements with very small payloads, smaller movements with extremely heavy parts, including entire vehicle frames, and specialized tasks such as surface prepping and painting, which added the elements of atmospheric contaminants.

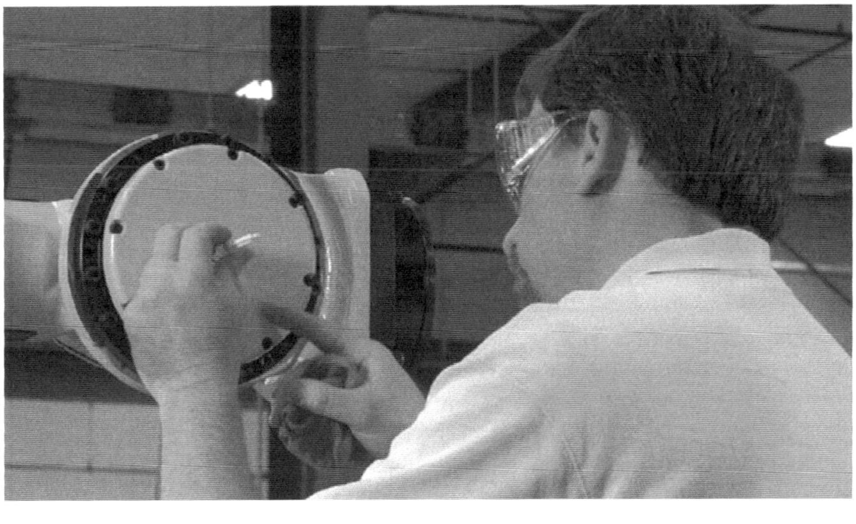

FIGURE 5: Robot Grease Sampling

Samples from each of the robots were taken in accordance with ASTM D7718 and submitted for analysis. A detailed test slate was used to determine the condition of the grease, contaminants and wear levels and correlated to runtimes for each sample taken. It became clear that some robot applications produced a rapid degradation rate of the grease, while others demonstrated the capacity for significant extension of life. In each case, the actual condition of the grease in the application, not generalized assumptions or conservative OEM guidance, was used to optimize grease use and replenishment intervals. This resulted in an overall net savings of $70,000 annually for the 300 robots in the study, with the additional benefit of gaining valuable insight into the condition of the robot joints.

CONCLUSION:

Grease utilization has historically been a rather inefficient process dominated by excessive labor activities, utilizing unnecessarily large quantities of grease, and often resulting in damage to equipment. Standard methods for obtaining representative samples, and performing valuable analysis with small sample quantities, has resulted in the availability of data that can be used to transition to condition-based greasing activities. At-machine solutions have been developed and can complement laboratory analysis activities to produce low-cost and effective grease component monitoring activities to improve sustainability, reduce labor and waste, and improve grease lubricated asset performance.

BIBLIOGRAPHY:

1. American Society for Testing and Materials, 2011. *ASTM D7718: Standard Practice for Obtaining In-Service Samples of Lubricating Grease.* West Conshohocken, PA, USA.

2. Air Transport Association of America (ATA), 2013. *ATA MSG-3 MSG-3 Volume 2 (Rotorcraft) Operator / Manufacturer Scheduled Maintenance Development.* Washington, DC: Airlines for America.

3. Møller, H., Esbensen, K. & Wurzbach, R., 2016. *Grease Sampling and Analysis For In-Service Condition Monitoring (CM) of Wind Turbine Blade Bearings.* Bilbao, Spain, s.n.

4. Pitard, F., 1989. *Pierre Gy's Sampling Theory and Sampling Practice, 2 Volumes.* Boca Raton, Florida: CRC Press, Inc..

5. Singer, T., Johnson, M. & Wurzbach, R., 2020. *Rotorcraft Lubrication Optimization through Grease Sampling and Analysis.* VTOL Forum 76. Virtual.

AUTHOR INDEX